Bioprocessing for Cell Based Therapies

Bioprocessing for Cell Based Therapies

Edited by Che J. Connon

Institute of Genetic Medicine,
Newcastle University,
Newcastle upon Tyne, UK

WILEY Blackwell

Registered Office
John Wiley & Sons, Ltd, The Atrium, Southern Gate, Chichester, West Sussex, PO19 8SQ, UK

Editorial Offices
9600 Garsington Road, Oxford, OX4 2DQ, UK
The Atrium, Southern Gate, Chichester, West Sussex, PO19 8SQ, UK
111 River Street, Hoboken, NJ 07030-5774, USA

For details of our global editorial offices, for customer services and for information about how to apply for permission to reuse the copyright material in this book please see our website at www.wiley.com/wiley-blackwell.

Library of Congress Cataloging-in-Publication data applied for:

9781118743416

A catalogue record for this book is available from the British Library.

Wiley also publishes its books in a variety of electronic formats. Some content that appears in print may not be available in electronic books.

Cover image: © Science Photo Library – STEVE GSCHMEISSNER/Gettyimages

Set in 10/12pt, Warnock Pro by SPi Global, Chennai, India

10 9 8 7 6 5 4 3 2 1

Contents

List of Contributors *xi*
Preface *xv*

1 Overview of the Cell Therapy Field *1*
Michael Whitaker, Lucy Foley and Stephen Ward
1.1 The Context of Cell Therapies and Their Manufacturing Challenges *1*
1.1.1 Regulation of Cell Therapies *4*
1.1.2 Manufacturing Challenges in Cell Therapy *5*
1.2 The Cell Therapy Landscape *5*
1.2.1 Licensed Cell Therapy Products *7*
1.2.2 Companies, Clinicians, Products and Procedures *8*
1.2.3 Cell Therapy Clinical Trials *8*
1.3 Operations in Cell Therapy Manufacture *11*
1.3.1 Cells for Cell Therapy Production *12*
1.3.1.1 Cell Source *12*
1.4 Upstream Processing of Cellular Therapies *13*
1.4.1 Cell Separation *13*
1.4.2 Cell Expansion *13*
1.4.3 Tissue Expansion *15*
1.4.4 Adherent Cell Expansion *15*
1.4.4.1 Multi-layer Reactors *15*
1.4.4.2 Hollow Fibre Reactors *16*
1.4.4.3 Scaffolds *17*
1.4.5 Suspension Cell Expansion *18*
1.4.5.1 Stirred Tank Bioreactors *18*
1.4.5.2 Rocking Platforms *19*
1.4.5.3 Perfusion Cell Culture *19*

1.4.6 Differentiation *19*
1.5 Downstream Processing of Cellular Therapies *20*
1.5.1 Harvest, Washing and Concentration *20*
1.5.1.1 Centrifugation *21*
1.5.1.2 Filtration *21*
1.5.2 Separation and Purification *22*
1.5.2.1 Centrifugation *22*
1.5.2.2 Magnetic Separation *24*
1.6 Formulation, Fill and Finish of Cellular Therapies *24*
1.6.1 Formulation *25*
1.6.2 Fill and Finish *25*
1.6.3 Preservation and Shipment *26*
1.7 Administration of the Cell Therapy to a Patient *27*
1.8 Cell Therapy Manufacturing Facilities of the Future *28*
1.8.1 Factory of the Future Requirements *31*
1.9 Conclusion *31*
References *32*

2 Structured Methodology for Process Development in Scalable Stirred Tank Bioreactors Platforms *35*
Huaqing Wang, Daniel Kehoe, Julie Murrell and Donghui Jing
2.1 Introduction *35*
2.2 Understanding the Engineering of the Stirred Tank Bioreactors *36*
2.2.1 Mixing Phenomena in Stirred Tank Bioreactors *37*
2.2.2 Understanding Oxygen Transfer Rate (k_La) with Different Sparging Methodologies *40*
2.2.3 Heat Transfer in STB (Minimum Volume, Sensor/Sensing Control) *42*
2.2.4 How to Choose a Microcarrier for Adherent Cells (hMSCs) *43*
2.3 Understanding the Biology of the Cells in Stirred Tank Bioreactors (STB) *45*
2.3.1 Cell Types (Adherent and Suspension Cells) *45*
2.3.2 Assays for Comparability, Nutrients, Senescence and Doubling Dime *46*
2.4 Process Development of Adherent Cells in STB Platforms *47*
2.4.1 Standard Comparison to Cell Factory *47*
2.4.2 Methodology for Screening of Microcarriers *49*

2.4.3 Process Development in Small-scale Bioreactors (3L) *50*
2.4.3.1 Cell Seeding Density *51*
2.4.3.2 Microcarrier Concentrations (Cells:Bead Ratio) *52*
2.4.3.3 Operation Ranges of pH and Dissolved Oxygen *53*
2.4.3.4 Feeding Strategy *54*
2.4.4 Process Development in Medium-scale Bioreactors (50L) *55*
2.4.5 Case Study for Expansion of Bone Marrow Derived MSCs in Stirred Tank Bioreactors *59*
2.5 Future Directions *60*
References *61*

3 The Effect of Scale-up on Cell Phenotype: Comparability Testing to Optimize Bioreactor Usage and Manufacturing Strategies *65*
Jason Hamilton and Bart Vaes
3.1 Introduction *65*
3.1.1 Cell Characterization in the Development Path *66*
3.1.2 The MultiStem® Allogeneic Cell Therapy Product: Mechanisms of Benefit and Target Cells Numbers *69*
3.2 Challenges in Cell Product Development *72*
3.2.1 Effect of Large-scale Expansion on Stem Cell Properties *72*
3.2.2 Serum-free and Xeno-free Media Development *74*
3.3 Stem Cell Characterization *75*
3.3.1 ISCT Requirements *75*
3.3.2 Potency Assays *77*
3.3.3 Omics Screens for Therapeutic Stem Cell Characterization *78*
3.4 Next-generation Stem Cell Development *80*
References *83*

4 The Scale-up of Human Mesenchymal Stem Cell Expansion and Recovery *91*
Thomas R. J. Heathman, Qasim A. Rafiq, Karen Coopman, Alvin W. Nienow and Christopher J. Hewitt
4.1 Introduction *91*
4.2 Scale-up or Scale-out *93*
4.3 Understanding the Small Scale *96*
4.4 Microcarrier Screening *103*

4.5 Spinner Flask Culture *108*
4.6 Large-scale Expansion in Conventional Stirred Tank Bioreactors *111*
4.7 Cell Recovery from Microcarriers *117*
4.8 Conclusions *120*
References *121*

5 Challenges of Scale-up of Cell Separation and Purification Techniques *127*
Marieke A. Hoeve, Paul A. De Sousa and Nicholas A. Willoughby
5.1 Introduction *127*
5.1.1 Cell Separation for Cell-based Therapeutics *127*
5.1.2 Separation Methodology Design *128*
5.1.3 Objective of this Chapter *128*
5.1.3.1 Cell Yield *129*
5.1.3.2 Standardisation *129*
5.1.3.3 Economical viability *129*
5.2 Scalable Cell Separation for Cell Therapy *130*
5.2.1 Label Requiring versus Label-free Separation *130*
5.2.2 Active versus Passive Method *133*
5.2.3 Isolated Purification (Including Off-the-Shelf) versus Embedded Integrated Process *133*
5.2.4 Low versus High Resolution *133*
5.2.5 Open versus Closed Systems *134*
5.2.6 Batch versus Continuous Separation *134*
5.3 Currently Developed Cell Separation Techniques *135*
5.3.1 Acoustophoresis *135*
5.3.2 Aqueous Two-Phase System (ATPS) *137*
5.3.3 Centrifugal Techniques *138*
5.3.3.1 Centrifugal Counterflow Elutriation (CCE) *138*
5.3.3.2 Centrifuge Systems with Integrated Filters *139*
5.3.4 Dielectrophoresis (DEP) *140*
5.3.5 Deterministic Lateral Displacement (DLD) *141*
5.3.6 Genetic Engineering *141*
5.3.7 Hydrodynamic Filtration (HDF) *142*
5.3.8 Immunoadsorption *142*
5.3.9 Immunomagnetic Cell Sorting *145*
5.3.10 Inertial Migration *145*

5.3.11 Magnetic Cell Sorting – Label-free *147*
5.3.11.1 Magnetophoretic Cell Separation *147*
5.3.11.2 Magnetic Solution-based Separation *148*
5.3.12 Microscale Vortices *150*
5.3.13 Normal Flow Filter (NFF) *150*
5.3.14 Optical – Label-free *151*
5.3.15 Tangential Flow Filters (TFF) *155*
5.3.16 Weir and Pillar *156*
5.4 Conclusion *157*
Acknowledgements *159*
References *159*

6 Fundamental Points to Consider in the Cryopreservation and Shipment of Cells for Human Application *167*
Glyn N. Stacey, Lyn Healy, Jennifer Man, Charles J. Hunt and John Morris
6.1 Introduction *167*
6.2 The Role of Cryoprotective Agents (CPA) *168*
6.3 Vitrification versus Cryopreservation *169*
6.4 Points to Consider in the Development of Cryopreservation Protocols *169*
6.4.1 General Considerations *169*
6.4.2 Cellular Characteristics and Selection of Appropriate CPAs and Cooling Protocols *170*
6.4.3 Key Events in Cryopreservation *172*
6.4.3.1 Ice Nucleation *174*
6.4.3.2 CPAs: Concentration and Composition *174*
6.4.3.3 Cooling Rate *174*
6.4.3.4 Storage of Cryopreserved Cells *176*
6.4.3.5 Thawing and Recovery of Frozen Cells *177*
6.5 Large-volume Freezing *178*
6.6 Cryopreservation as Part of Manufacturing Processes *179*
6.6.1 Containers for Cryopreserved Cells *179*
6.6.2 Controlled Rate Freezers and Storage Systems *179*
6.6.3 The Cold-chain: Challenges and Solutions *180*
6.6.4 Biobanking and Regulatory Requirements *181*
6.7 Conclusions *181*
Acknowledgements *182*
References *182*

7 Short-term Storage of Cells for Application in Cell-based Therapies *187*
Stephen Swioklo and Che J. Connon
7.1 Introduction *187*
7.1.1 Advances in Cell-based Therapies *188*
7.1.2 The Logistical Landscape for CTPs and the Requirement for Short-term Storage of Cells *188*
7.2 Hypothermia and Mammalian Cell Storage *193*
7.2.1 Hypothermic Storage of Mesenchymal Stem Cells (MSCs) *194*
7.2.2 Optimal Temperature for Cell Storage *202*
7.3 The Application of Hypothermic Storage in Cell-based Therapies *204*
7.4 Concluding Remarks *205*
References *205*

8 Cell Therapy in Practice *211*
Gustavo S. Figueiredo, Julian R. De Havilland, Majlinda Lako and Francisco C. Figueiredo
8.1 Introduction *211*
8.2 The Classification of ATMPs *212*
8.3 European Regulations *216*
8.3.1 Hospital Exemption (HE) and "Specials" Manufacturing *219*
8.3.2 Orphan Medicinal Product Designation *220*
8.3.3 Committee for Advanced Therapies *221*
8.3.4 Good Manufacturing Practice (GMP) *221*
8.3.5 European Union Tissue and Cells Directives (EUTCD) *223*
8.4 ATMP Case Study: Autologous Limbal Stem Cell Therapy: the Newcastle Experience *224*
8.5 Conclusion *233*
References *234*

Index *237*

List of Contributors

Che J. Connon
Institute of Genetic Medicine
Newcastle University
Newcastle upon Tyne
UK

Karen Coopman
Centre for Biological Engineering
Loughborough University
Leicestershire
UK

Julian R. De Havilland
Cellular Therapies Facility
Newcastle University
Newcastle upon Tyne
UK

Paul A. De Sousa
Centre for Clinical Brain Sciences and
MRC Centre for Regenerative Medicine
University of Edinburgh
Edinburgh
UK

Gustavo S. Figueiredo
Department of Ophthalmology
Royal Victoria Infirmary
Newcastle upon Tyne
UK

Institute of Genetic Medicine
Newcastle University
Newcastle upon Tyne
UK

Francisco C. Figueiredo
Department of Ophthalmology
Royal Victoria Infirmary
Newcastle upon Tyne
UK

Institute of Genetic Medicine
Newcastle University
Newcastle upon Tyne
UK

Lucy Foley
Centre for Process Innovation – Biologics
Darlington
UK

Jason Hamilton
Athersys Inc
Cleveland, OH
USA

Lyn Healy
Haematopoietic Stem Cell Laboratory
The Francis Crick Institute
London
UK

Thomas R. J. Heathman
Centre for Biological Engineering
Loughborough University
Leicestershire
UK

Christopher J. Hewitt
Centre for Biological Engineering
Loughborough University
Leicestershire
UK

Aston Medical Research Institue
School of Life and Health Sciences
Aston University
Birmingham
UK

Marieke A. Hoeve
Centre for Clinical Brain Sciences and
MRC Centre for Regenerative Medicine
University of Edinburgh
Edinburgh
UK

Charles J. Hunt
UK Stem Cell Bank
Advanced Therapies Division
National Institute for Biological Standards and Control
Hertfordshire
UK

Donghui Jing
HaoLing Cell Technologies Corporation
TEDA
Tianjin
P.R.China

Daniel Kehoe
Stem Cell Bioprocessing Group
Process Solutions
EMD Millipore Corporation
Bedford
USA

Majlinda Lako
Institute of Genetic Medicine
Newcastle University
Newcastle upon Tyne
UK

Jennifer Man
UK Stem Cell Bank
Advanced Therapies Division
National Institute for Biological Standards and Control
Hertfordshire
UK

John Morris
Asymptote Ltd
St John's Innovation Centre
Cambridge
UK

Julie Murrell
Stem Cell Bioprocessing Group
Process Solutions
EMD Millipore Corporation
Bedford,
USA

Alvin W. Nienow
Centre for Biological Engineering
Loughborough University
Leicestershire
UK

Centre for Bioprocess Engineering
University of Birmingham
UK

Qasim A. Rafiq
Centre for Biological Engineering
Loughborough University
Leicestershire
UK

Aston Medical Research Institute
School of Life and Health Sciences
Aston University
Birmingham
UK

Glyn N. Stacey
UK Stem Cell Bank
Advanced Therapies Division
National Institute for Biological Standards and Control
Hertsfordshire
UK

Stephen Swioklo
Institute of Genetic Medicine
Newcastle University
Newcastle upon Tyne
UK

Bart Vaes
ReGenesys BVBA
Heverlee
Belgium

Huaqing Wang
Department of Medical Oncology
Nankai University Cancer Hospital
Tianjin Union Medical Center
Tianjin
P.R.China

Stephen Ward
Cell and Gene Therapy Catapult
Guy's Hospital
London
UK

Michael Whitaker
Institute of Cell and Molecular Sciences
Newcastle University
Newcastle upon Tyne
UK

Nicholas A. Willoughby
Institute of Biological Chemistry
Biophysics and Bioengineering
School of Engineering and Physical Sciences
Heriot-Watt University
Edinburgh
UK

Preface

Cell-based therapy is an exciting and rapidly developing field of medicine for the 21st century. However, this new paradigm brings new challenges in regulation and production, which requires cross-disciplinary approaches and greater collaboration between clinicians, academics and industrial scientists. This new alliance is reflected in the range of disciplines from which the distinguished contributing authors have been drawn to create this book, including Industry (EMD Millipore Corp., Athersys Inc., ReGenesys, Asymptote Ltd), Governmental (Process Innovation Centre – Biologics, Cell Therapy Catapult, UK Stem Cell Bank) and Academic Institutions (The Francis Crick Institute, and Newcastle, Birmingham, Loughborough, Aston, Edinburgh and Heriot-Watt universities). Together they represent the cutting edge, industrially focused frontline of bioprocessing for cell-based therapies.

The book is composed of eight chapters beginning with an introductory chapter that sets out the history of cell therapy leading up to the current and future challenges associated with the manufacture and distribution of this relatively new class of therapeutic modality. This includes a detailed discussion on the advantages and disadvantages of the one-one and one-many approaches in cell-based therapy. The book then moves into a series of chapters focusing on recent technical developments in bioprocessing for cell-based therapies. Chapter 2 discusses the use of stirred tank bioreactors for hMSC cultivation, specifically engineering requirements such as mixing phenomena, oxygen and heat transfer rates, metabolic demands of the cells and particular problems associated with adherent cells. Chapter 3 follows this theme by focusing on the important topic of cell characterization during scale-up of hMSCs. We are informed that to guarantee product consistency there needs to be uniformity in identity, maintenance of a unique product character and crucially demonstration of a consistent and therapeutically relevant product potency. Chapter 4 also focuses on hMSC culture, asking if we need to consider scale-up or scale-out. This chapter, then proceeds to highlight the technical difficulties involved in moving bioprocessing research from the lab-scale to large-scale, covering the challenges associated with choice of microcarriers and spinner flasks, then removing cells from chosen microcarriers. Chapter 5 continues by describing the importance of cell separation for cell-based therapies with an in-depth focus on the current methods for mammalian cell separation and their suitability for application in large-scale cell purification for clinical application. We then move on to Chapters 6 and 7, which discuss the importance of storage and transport of the cellular products. Chapter 6 focuses on cryopreservation, how to limit cell damage, large volume freezing, approaches to biobanking and regulatory requirements. Chapter 7 takes an interesting departure from cryopreservation, instead focusing on cold or hypothermic cell preservation, in which the logistics of short-term storage are described as are the use of hydrogels as a novel storage medium. The final chapter (Chapter 8) focuses on where the advancements in bioprocessing end up, such as in the clinical application of cell-based therapies. This chapter includes a case study describing the development of a new therapy against the background of a rapidly changing regulatory environment and the challenges this poses, before bringing the therapy successfully to the clinic.

1

Overview of the Cell Therapy Field

Michael Whitaker[1], *Lucy Foley*[2] *and Stephen Ward*[3]

[1] *Institute of Cell and Molecular Sciences, Newcastle University, Newcastle upon Tyne, UK*
[2] *Centre for Process Innovation – Biologics, Darlington, UK*
[3] *Cell and Gene Therapy Catapult, Guy's Hospital, London, UK*

1.1 The Context of Cell Therapies and Their Manufacturing Challenges

Cell therapies are not new. The first cell therapy was the transplant of bone marrow stem cells for patients with leukaemia in the late 1970s (Thomas et al., 1975). Over the next 20 years bone marrow stem cell transplants were adapted and adopted for bone marrow cancers and extended into other clinical indications, for example, inherited immunodeficiency. Bone marrow stem cell transplantation is now a routine clinical procedure for multiple indications.

The example of bone marrow stem cell therapies illustrates several of the characteristics that define cellular therapies more broadly. They were developed for less prevalent indications. They are allogeneic (see Box 1.1) one donor-one recipient therapies that were developed wholly by clinicians in a hospital context. They require interventional clinical procedures for administration of the therapy. They led to widespread clinician-led adoption by the clinical community through global clinician networks. They defined an approach to the **safety** of cell therapies based on risk and benefit to patients. They encouraged the development of cell processing expertise within hospitals and in many ways provided a basis of skills and expertise for clinicians to facilitate the development of other cell

Bioprocessing for Cell Based Therapies, First Edition. Edited by Che J. Connon.

therapies (Foley and Whitaker, 2012). Though led by clinicians, the development and widespread adoption of bone marrow stem cell therapies was facilitated by companies who provided high-value goods and services to help manufacture and deliver the therapies in a hospital context. Around the turn of the millennium, two cell therapy products developed by companies were the first cell-based therapies to be approved by the US Food and Drug Administration (FDA). Apligraf and Dermagraft were competing skin-equivalent products designed to improve the healing of wounds and burns (Kemp, 2006).

These two therapies illustrate some of the characteristics of company-led approaches to cell therapies. They are allogeneic, one donor-many recipient cell therapy products that are manufactured at scale for prevalent indications. They do not require complex clinical procedures (Foley and Whitaker, 2012).

Apligraf and Dermagraft initially failed in the market (Lysaght and Hazlehurst, 2004). One key reason was cost of goods: the products were manufactured manually and had a short shelf life – two aspects

Box 1.1 Cell Therapy Definitions

Autologous

The patient's cells are the cells used in the therapy. No immune response is expected.

Allogeneic

The cell source is different to the patient receiving the cell therapy. There is a possibility of an immune response.

One to one

The cells used in the therapeutic dose are only of sufficient quantity to treat one patient; these treatments can be autologous or allogeneic.

One to many

The cells used are amplified to a scale able to treat many patients; these treatments can only be allogeneic.

of manufacture and distribution that are not well suited to prevalent indications. A contributing factor was cost relative to existing treatments, despite improved efficacy (see Box 1.2).

Box 1.2 Dermagraft and Apligraf – a roller coaster of investment, manufacturing costs and reimbursement

In the 1990s, Advanced Tissue Sciences invested around \$300m to develop Dermagraft and Transcyte for the treatment of diabetic foot ulcers. In 2000, ATS formed a marketing partnership with Smith and Nephew, a global leader in wound care products. Dermagraft was approved by the FDA in 2001. In 2002, ATS filed for bankruptcy. In 2003, Smith and Nephew purchased ATS from bankruptcy and continued with manufacturing and sales. Smith and Nephew ceased production in 2005. In 2006, Advanced Biohealing purchased the Smith and Nephew manufacturing assets for an undisclosed amount (Jones, 2011), presumably at a value destroying discount, and in 2007 resumed manufacture, with a sales/reimbursement model that led to \$147m sales in 2010. In 2011, Shire bought ABH for \$750m (Smith, 2014). In 2013, Dermagraft assets were declared at \$683m on Shire's balance sheet and 9 month losses for Dermagraft were \$324m (Reporter, 2014).

Organogenesis was the first to receive FDA approval for a living, allogeneic, cell-based product (Apligraf). They were successful in securing a marketing agreement with Novartis in 1996 (Connolly, 2002a). However, the cost of producing Apligraf was too high and in 2002 Organogenesis filed for bankruptcy and terminated its marketing agreement with Novartis. A short-term deal with Novartis and company restructure (Connolly, 2002b) today means that Organogenesis develops, manufactures and markets its own products.

In early 2014, Organogenesis acquired Dermagraft from Shire, with a promise of a \$300m payment based on future sales, but without accepting liability for the ongoing Department of Justice investigation into ABH sales and marketing practices (GenEngNews, 2014). Later that year, Medicare altered reimbursement rules (Carroll, 2013), suggesting that the \$1,500 cost of Dermagraft would be reimbursed at a maximum of \$840. Dermagraft is a very effective treatment for diabetic ulcers, but costs and reimbursement routes may prevent it reaching patients.

1.1.1 Regulation of Cell Therapies

Neither bone marrow stem cell transplantation nor the first two marketed cellular products were regulated as cell therapy products now are. A key question in the current regulatory landscape is whether cells are substantially manipulated before administration to the patient. Minimally manipulated cells, for example, using aseptic separation or enrichment, are governed by the same regulations that apply to any cell or tissue taken from a patient. Therapeutics that involve any more substantial manipulation including expanding cell numbers are now governed by different and more stringent rules akin to those used in the regulation of other medicinal products such as small molecule pharmaceuticals and biologics. These rules require that quality, safety and efficacy are demonstrated to the satisfaction of the regulators, both in order to undertake the clinical trials and for authorization as an approved medicinal product if the trials are successful. The rules include a requirement to show that the product has been manufactured according to Good Manufacturing Practice. So products similar to Apligraf and Dermagraft now require these new authorizations, while bone marrow stem cell transplantation with its minimal manipulation before administration does not and is overseen for Good Clinical Practice by organizations such as JACIE in Europe. In the US, the FDA oversees Good Clinical Practice for bone marrow transplantation, but this is separate from Biologics manufacture; in the US, the term "Biologics" encompasses cell therapies and the more traditional biopharmaceuticals (Oancea et al., 2012).

The uncertainties that preceded the introduction of the new regulations and the costs in time and money that were required for compliance with the new regulations led to a pause in the development of cell therapies, above all in the US and Europe. Nonetheless it is notable that both clinician- and company-led cell therapies have adapted to the regulatory change. As an example of the former, 12 of the 26 cell therapy manufacturing facilities in the UK are now accredited for Good Manufacturing Practice-compliant manufacture of cellular products (Foley et al., 2012). It should be acknowledged that the new regulations for cell therapies are very similar to those for biologics, and so are well understood by the pharmaceutical sector. They do however still pose a substantial manufacturing challenge, since it is the cells themselves, not a biotherapeutic product produced by cells, that are the medicinal product (the ATMP: Advanced Therapeutic Medicinal Product).

1.1.2 Manufacturing Challenges in Cell Therapy

The key raw material for cell therapy manufacture is a cell type obtained from a human source. A key distinction between cell therapies is whether the cell type or its differentiated or otherwise modified derivatives are destined for a single patient or for many patients. A second difference is whether the cells of origin are administered to the patient from whom they are taken. If they are, then they have the genetic identity of the patient and the therapy is autologous. If they are not, then the cells are genetically distinct from the patient recipient and the therapy is allogeneic. For the most part, allogeneic therapies are one to many, while autologous therapies are one to one, though there are examples of one to one allogeneic therapies in which a single patient is treated with cells from a single genetically distinct donor (see Box 1.1).

The manufacture of one to many cell therapies closely resembles the manufacture of biopharmaceuticals (Figure 1.1). Cells from the donor are grown, separated and characterized to make a master cell bank (Box 1.3). Working cell banks can be derived from the master cell bank and used to manufacture patient doses, as for biopharmaceuticals. However, there are important and challenging differences (see Figure 1.1).

We have already briefly mentioned one difference. It is the cells themselves, not their products that are the therapy; as is often remarked, in biopharmaceutical manufacture, one throws away the cells, while in cell therapy manufacture, one throws away the medium. A second crucial difference is that biopharmaceutical manufacture relies on a few standard cell types in standard media; cell therapy manufacture is bespoke to each therapy and does not have the benefit of well-developed platform technologies, for the time being at least.

One to one cell therapy manufacture uses many of the underlying processes and principles of biopharmaceutical manufacture, but is markedly different in scale and separation technologies. Each dose is manufactured for a single patient from a single donor and multiple doses must be manufactured in parallel (Box 1.4).

1.2 The Cell Therapy Landscape

The REMEDiE project has identified around 700 companies that are working in regenerative medicine products or services worldwide, with the large majority located in the US or Europe. More than 90%

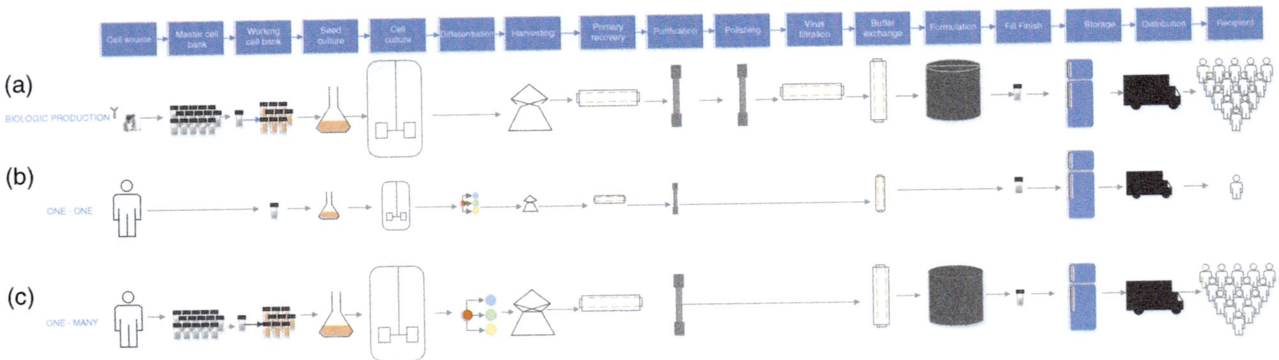

Figure 1.1 Process flow diagram of the sequence of unit operations used in biopharmaceutical manufacture and their relative scales: (a) general process used for the manufacture of monoclonal antibodies; (b) manufacturing for a one on one therapy; and (c) manufacturing process for a one to many cell therapy process.

Box 1.3 How many cells? The scale of manufacture

Mason and Dunhill (2009) and Simaria (2014) have provided some estimates of the numbers of cells that may need to be manufactured. Individual cell doses range from 10^5–10^8. A one to one therapy such as cartilage repair needs a dose of around 10^7 cells and can scale to around 10^4 patients. Treatment for heart failure (one to many) may require a 10^9 dose in 10^7 patients, that is, the manufacture of 10^{16} cells. If cells weigh around 10^{-8} kg, this is around 100,000 metric tonnes: a manufacturing challenge.

Box 1.4 Major manufacturing challenges

Bespoke laboratory based manufacturing processes for early clinical trials may not scale economically to satisfy demand related to disease prevalence. Thus there is a need for early thinking in process development:

- developing robust, replicable processes that can be scaled up or out
- moving early lab-based processes to GMP-compliant processes, materials and equipment; and
- incorporating supply chain and clinical delivery in process development.

Funding process development is a challenge given current public and private funding frameworks.

of these companies are SMEs. Regenerative medicine is defined as including cell therapies, but not exclusively, so it is likely that the number of companies working specifically on cell therapies is somewhat smaller.

1.2.1 Licensed Cell Therapy Products

There are currently 11 cell therapies licensed by the FDA (2014). Of these, five are cord blood derived haematopoietic progenitor cells (HPCs), three are based on fibroblasts or keratinocytes or both, two are chondrocyte-derived and one is a modified dendritic cell. In Europe, only two cell therapies have so far been approved by the EMA and both are chondrocyte-derived (Tozer, 2011). The HPCs are allogeneic one to one therapies for bone marrow disorders delivered by clinicians; of the three skin cell therapies, two are allogeneic and all are marketed by companies; the chondrocyte therapies are autologous and also marketed by companies. Licensed therapies in Europe and the US are described in Figure 1.2.

1.2.2 Companies, Clinicians, Products and Procedures

In analyzing the cell therapy landscape we have found it useful to distinguish between a cell therapy that can be readily administered to a patient and another that requires a more complex clinical intervention to deliver it to the site of choice. The former we have called a product because it is closest to an off-the-shelf drug or biopharmaceutical. We have classified the latter therapies as requiring a procedure (Foley and Whitaker, 2012). From clinical trial data, the tendency is seen to be for procedure-based therapies to be set up and delivered by clinicians, while product-focused cell therapies tend to be developed by companies. Mapping of the existing licensed therapies shows that 10 involve a clinical procedure and 3 a product. Companies have developed all the licensed products, while the procedures have been developed equally by clinicians and by companies (Figure 1.2).

1.2.3 Cell Therapy Clinical Trials

In an exhaustive analysis of worldwide clinical trials databases, around 1,000 cell therapy trials were found that were not investigating established cell therapies for an established indication (Li et al., 2014). Of these 1,000 trials, just over 400 studied mesenchymal stem cells and an equal number used haematopoietic stem cells. The number of trials based on other stem cell types was 208, including only 6 that used embryonic stem cells, the remainder being somatic stem cells. Around 600 trials used autologous cells and 300 allogeneic cells as a therapeutic agent; around 100 trials involved stimulating endogenous cells with a non-cellular therapeutic.

These figures are broadly comparable to an analysis undertaken in 2011 (Foley and Whitaker, 2012) that used a sampling algorithm using data only from www.clinicaltrials.gov, which might be expected to show a bias towards US and European cell therapy trials. Using the same methodology, we have now analyzed the trials database again; however, this time including additional trials that were registered up

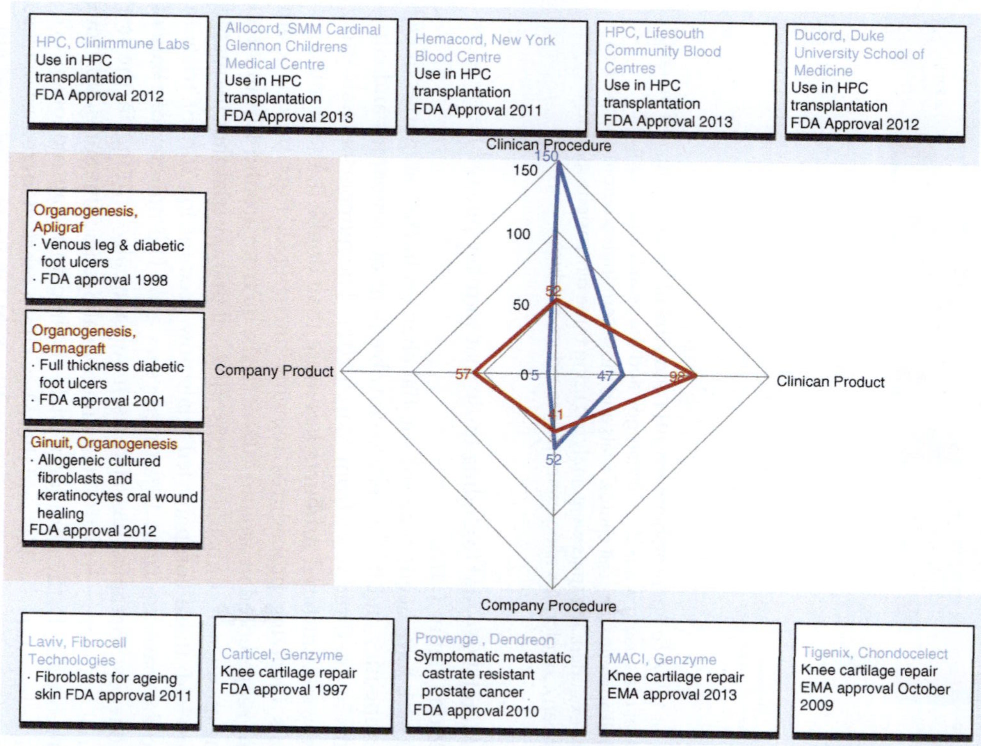

Figure 1.2 Heuristic classification of a selection of cell therapies to the three dimensions. Therapies shown in red are one to many, and therapies in blue are one to one. Numbers are our estimate of all stem cell trials in the clinical trials database that met our criteria. Therapies in boxes have received a market authorization. Abbreviations: EMA, European Medicines Agency; FDA, US Food and Drug Administration; HPC, human progenitor cells.

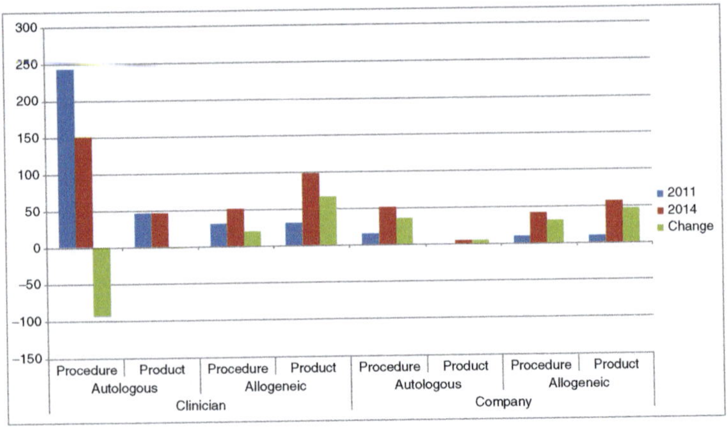

Figure 1.3 A comparison between 2011 and 2014 of the change in cell therapy clinical trials using www.clinicaltrials.gov and the search term "stem cells". Hematopoietic stem cell clinical trials and those terminated before completion were excluded. The trials were analyzed using three dimensions, clinician-company, autologous-allogeneic and procedure-product.

to March 2014 and excluding those that were terminated before completion (Figure 1.3).

There have been some substantial changes in the categories of cell therapeutics and in the relative involvement of companies and clinicians in the three years from March 2011. The number of trials using easily administered products has more than doubled; the number of trials involving procedures remains unchanged, thus products are now represented in around 40% of trials, up from around 17% in 2011. The number of companies sponsoring cell therapy trials has increased four-fold: these company-led trials now account for 30% of trials in the database, up from 10% three years ago. The number of trials of allogeneic therapies has also risen substantially, tripling in the last three years; now half the trials involve allogeneic therapies, up from 20%. Strikingly, the number of trials of clinician-led autologous procedures has fallen by 40% and there have been marked increases in trials of allogeneic therapies sponsored by both clinicians and companies (Figure 1.3).

The analyses of cell therapy clinical trials presented here and previously (Foley and Whitaker, 2012; Li et al, 2014) exclude cellular therapies for cancer. Cellular cancer therapies broadly involve the modification of immune cells to sensitize them to neoplasms. They are in general one-to-one therapies. The FDA has licensed 11 products to date, including Provenge for prostate cancer (Figure 1.2). Fifteen of the 41 stem cell trials identified in the UK by the Cell Therapy Catapult (an organization fostering and supporting stem cell therapies, including their manufacture) were immune cell therapies, mostly for cancer (Mount, 2014).

1.3 Operations in Cell Therapy Manufacture

The supply chain for cell therapies is complex. It begins with isolation of a cell source from the donor and terminates in delivery of the cell dose to the patient. Processing steps for the two key types of cell therapy, one to one and one to many follow similar paths, though the scale of operation is different (Figure 1.1).

Once the cells have been sourced from a donor or working cell bank, the manufacturing processes for cell therapies involve a number of unit operations (Figure 1.4). First there are unit operations to multiply the number of cells to obtain the quantity required for therapeutic dose(s) and perhaps to differentiate them; these unit operations are described as upstream processing. Secondly, the cells are purified out of the culture media used to grow them so that they are suitable for administration to a patient; the unit operations employed to do this are described as downstream processing. Prior to shipment, cell therapies must be formulated and packaged into a therapeutic dose, this being described as formulation, fill and finish. Finally, at the site where the cell therapy is to be administered, there will be some reformulation and preparation. The next section will outline the unit operations used in the processing of cell therapies from the cell source through to delivery to the patient.

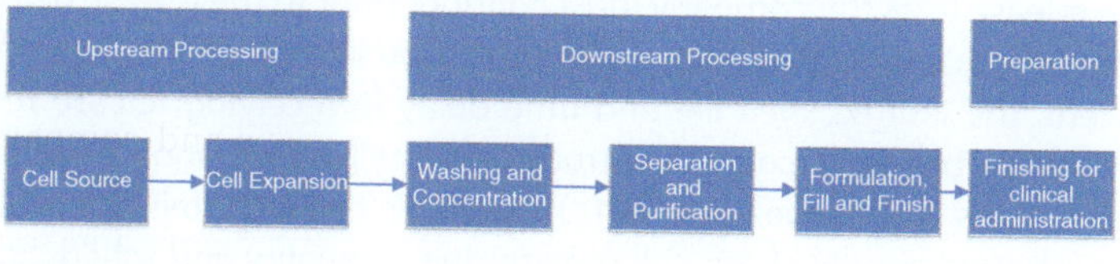

Figure 1.4 Operations performed in upstream and downstream processing of cell therapies.

1.3.1 Cells for Cell Therapy Production

1.3.1.1 Cell Source

Initial cell samples are retrieved from a human volunteer or patient. The tissue source of cells varies. Cell therapies in development illustrate retrieval from a wide variety of sources; biopsies of the skin, hair follicles and eye are common, as are retrievals from peripheral blood and bone marrow (Hipp and Atala, 2008). Procurement regulations and guidelines require ethical approval and patient consent for the biopsy and the uses to which cells are put. In particular, cells that will result in a commercial product require specific consent for this purpose (Box 1.5).

A major difference between one to one and one to many therapies is that cells that are retrieved for one to one therapies are often limited in supply, so a large proportion of the sample will be used as the therapeutic dose. Furthermore, the inward supply chain (shipment of a patients cells or biopsy) for one to one cell therapies is more time critical and requires greater attention to scheduling than one to many cell

Box 1.5 Chondrocelect

Chondrocelect is a Tigenix NV (Leuven) product. It is the first ATMP has been approved by the EMA (Pearce et al., 2014). It is a one to one treatment for cartilage lesions and is manufactured by Pharmacell in the Netherlands. The manufacturing process consists of a biopsy being retrieved from the patient in a hospital setting. The biopsy is then shipped to the Netherlands where the cells are expanded and the final product formulated. This must then be shipped back to the patient where it is implanted into the knee in a hospital. This means two hospital appointments must be scheduled, the timing of the latter dependent on how well the patient's cells grow. Furthermore, product shelf life is a time limiting factor in the shipment. This is further complicated by the release tests the company must conduct prior to the release of the final product, which take time and impact product shelf life. To some extent, the limited shelf life and time delay in receiving release test results has been overcome by introducing a cell preservation step into the process (Vanlauwe et al., 2014). However, the uncertainty around the inbound supply chain, cell growth due to quality and scheduling hospital times still remains.

therapies (Wei Teng et al., 2014). Cells retrieved for one to many cell therapies must be proliferative enough to generate a master cell bank and further working cell banks from which many patients can be treated. As a consequence of these differences, one to one cell therapies usually require little or no long-term storage of either the donated cells or the therapeutic product. For one to many cell therapies, preservation and storage whilst maintaining the ability to proliferate is mandatory. Proliferative capacity in itself is not sufficient: master cell banks and working cell banks must proliferate without phenotypic or genotypic variation and consistently.

1.4 Upstream Processing of Cellular Therapies

1.4.1 Cell Separation

After retrieval of cells from the donor, it is often necessary to separate out the required cells prior to their expansion. Cell labelling and magnetic separation or density gradient separation are usually applied to complete this (Box 1.6). Cell separation may form part of a procedure to purify cells for direct administration to the patient. In this case, the cells are considered to be minimally manipulated.

1.4.2 Cell Expansion

When the retrieved cell sample does not contain enough of the required cells for a therapeutic dose, expansion of the isolated cell population is required. This expansion step means that the cells are no

Box 1.6 Sepax

Biosafe have a centrifugal cell processing system called the Sepax. This system is a fully automated cell separation device that is primarily used for the separation of cord blood, bone marrow and peripheral blood. It is the blood processing industry standard and is GMP, GLP and Point of Care compliant. The Sepax technology is also used in cell separation, washing and concentration of cell therapies and is currently a part of the manufacturing process for cell therapies in clinical trials, for example the Cell Medicas Cytovir CMV product which is in stage 3 clinical trials.

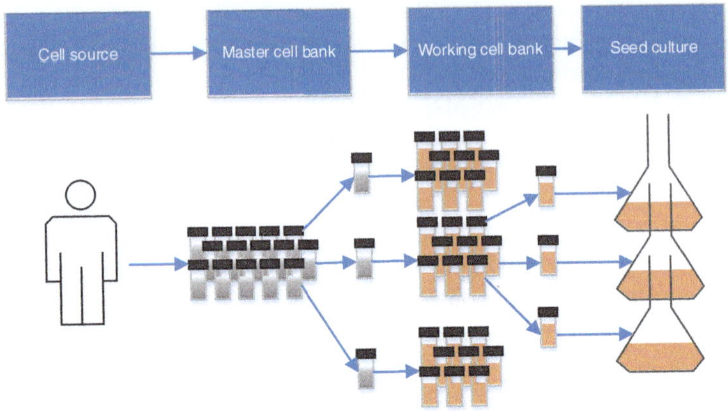

Figure 1.5 Cell banks are created for sustained manufacture of a cell therapy. From a master cell bank, working cell banks are developed which are then used in the manufacturing process.

longer minimally manipulated, so more complex manufacturing regulations apply. In terms of scale, for one to one therapies, the cells must be expanded sufficiently to provide a sufficient therapeutic dose to treat the individual patient. However, for one to many cell therapies, expansion is at multiple levels. Firstly, the cells must be expanded to create a master cell bank that is stored. Each vial from this cell bank then creates a working cell bank from which a single vial can be expanded to create enough cells to treat multiple patients (Figure 1.5). The proliferative capacity of the working cell bank must be sufficient to ensure genomic and phenotypic stability, not only at the point of withdrawal of an aliquot for manufacture, but also during any further proliferative steps during the manufacturing process itself.

We will discuss here three types of cell expansion: tissue expansion, adherent expansion and suspension expansion. Chapter 2 will address the challenges of industrial-scale cell expansion.

In all of the technologies used for cell expansion for cell therapies there has been a ready adoption of disposable materials that come into contact with cells. This limits the risks of contamination but also provides a more cost-effective option when compared with the cost of cleaning and validation of cleaning processes. However, the use of disposables brings the challenge of integrating manufacturing products supplied by different vendors into a seamless manufacturing process.

1.4.3 Tissue Expansion

For skin and for corneal biopsies (as described in Chapter 8) expansion occurs on a flat surface and the tissue sample, given the correct culture environment and media, will expand outwards. This process of expansion is limited by the tissue's ability to expand outwards as it would in its natural environment. This sort of expansion is therefore limited in scalability. Production of cell therapies in this form is thus limited to small surface area one to one treatments. One to many cell products require manufacture at scale and for adherent cell expansion this involves manufacture in parallel (scale-out); ideally, automation is used in order to reduce the otherwise large manpower costs of scale and associated processing risks that arise in manual manufacture at scale.

1.4.4 Adherent Cell Expansion

For most human single-cell expansion to date, the environment found to be most suitable is a flat surface, though this environment does not fully mimic the 3D microenvironment of much of the human body. Adherent cell suspension is the technique in which cells are expanded on a surface, their adherence to which provides stability throughout their expansion. Adherent culture is the most common type of cell expansion for cell therapies and can be commended as a technique that has enabled many cell therapies to enter clinical trials; it is the most widely used (in its simplest form) in discovery laboratories. The simplest type of adherent cell culture is through the use of plastic ware, for example Petri dishes or T-flasks. Cells are grown on the surface of a T-flask in an incubator that controls the environment. When the cells reach a specific cell density they are re-suspended and the flask contents are divided into further flasks for continued growth; this is termed a "passage". There are many advantages to this technique in that it is easy to implement and is well documented. However, the use of T-flasks limits scalability and is usually a manual and open process; thus it is subject to operator variation and the risk of contamination is high unless the processes can be automated.

1.4.4.1 Multi-layer Reactors

The use of T-flasks limits the degree of scalability, since to achieve scale, more and more T-flasks must be used with costs and risks rising rapidly with scale. One approach has been to automate the production

Box 1.7 Bioreactor type comparison

Work done at UCL (Simaria et al., 2014) looks at the number of cells required per dose for one to many cell therapies and the scale of production. A model has been developed that looks at the maximum productivity of different types of reactor used for cell expansion. Based on a patient dose of 10^9 cells the maximum feasibly achievable in T-flasks (80 500 mL flasks) is 0.9 of a dose, in an 80 ten-layer reactor it is 11 and in 8 2,000 L bioreactors using micro-carriers it is 4,708. This technology shift drastically reduces the cost of goods from $49 per million cells in T-flasks and $15 per million cells in multilayer reactors to 70 cents per million cells in bioreactors using microcarrier technologies.

of therapies in T-flasks; systems such as Sartorius Stedims Compact SelecT can process up to 90 T-flasks in one process run. Another approach has been through the development and use of "multi-layer reactors" (as described in Chapter 2). Multi-layer reactors provide the 2D environment required for adherent cells to grow by stacking many surfaces on top of one another within a single container. In essence, they may be thought of as many T-flasks surfaces stacked together in a single vessel. This increases the footprint productivity of a manufacturing facility. Many different types of multi-layer reactor are available, ranging from the most basic to compact versions and automated closed versions that re-circulate and refresh media (Box 1.7).

Multilayer reactors go some way to overcoming the scalability of adherent culture and are adequate for the manufacture of cell therapies where the dose size and target patient population are relatively small.

1.4.4.2 Hollow Fibre Reactors

Hollow fibre reactors are also used for adherent cell culture. For a given volume of container they maximize the surface area required for cells to adhere and grow whilst also maximizing the nutrient and oxygen exchange to the cells. There are many hollow fibre reactors on the market but all follow the same principles. The nature of hollow fibre bioreactors means that they are continuous, closed systems generally supported by a level of automation that facilitates media flow and mass transfer. However, as many of these bioreactors were developed for products that are secreted from cells, process

Box 1.8 Manufacture of Multistem

Multistem is a cell therapy product under development by Athersys. It is an allogeneic therapy which comprises of mesenchymal progenitor cells (MPCs) and is in clinical trials to treat inflammatory and immune, cardiovascular and neurological conditions (www.athersys.com). Early development of Multistem was completed using 2D culture such as T-flasks, and current GMP manufacture for clinical trials is completed in multilayer reactors by their contractor Lonza Biologics, Walkersville. In 2014, Athersys announced a partnership agreement with Terumo to develop their production process for MPCs using Terumo's hollow fibre reactor system, the Quantum. The Quantum is a closed, automated hollow fibre reactor that streamlines the cell culture process and reduces labour costs by up to 80% (www.terumobct.com). It is believed that both manufacturing routes will proceed in order to meet the wide variety of indications intended for this product. However, the Quantum will facilitate a more robust process due its closed, automated design.

optimization has been required to improve the yield of cells: in our hands, high yields are now possible. Whilst hollow fibre reactors are limited in their scalability (Hambor, 2012) they are nonetheless used successfully in a commercial setting for the manufacture of Multistem (Box 1.8) (Chapter 3). However there currently remains a strong desire to develop to suspension cell culture where the productivity per footprint, automation and process control are potentially much greater.

1.4.4.3 Scaffolds

Scaffold technology can also be used to grow cells for therapy. This is more usual for cells that require a 3D structure for administration, for example bone replacement. Scaffolds provide a micro-environment that is favourable for the cells to expand on and provide a structure through which they can proliferate. Cell therapy scaffolds can then be administered directly to the patient to provide a structure that will either remain indefinitely, such as Genzyme's MACI (matrix applied characterized autologous cultured chondrocytes) (De Bie, 2007) or degrade over time, such as the polymer scaffolds developed by Kanczler et al. (2008). There are many types of scaffolds under investigation. These can be from natural sources such as proteins and polysaccharides or synthetic sources such as polymers, peptides and ceramics (Willerth and Sakiyama-Elbert, 2008). The next generation of surfaces and environments need to be developed to mimic both developmental and disease niches. This should generate cells of enhanced functionality and increase yields.

1.4.5 Suspension Cell Expansion

Suspension cell expansion (as described in Chapters 2 and 4) is the method of choice for the production of biopharmaceutical drugs where the cells are used as factories to produce a therapeutic protein. Suspension cell culture has been developed over many years for the relatively few cell lines used in the manufacture of biopharmaceuticals. Suspension culture maximizes cell yields produced on a small footprint, using cell lines adapted to suspension culture. However, in the production of cell therapies, where the cell itself is the product, the ability of cells derived from normal tissues to proliferate and remain a viable therapy in suspension culture is limited. Thus the optimal conditions for human cells used in cell therapies growing in suspension culture are still very much in development (Kirouac and Zandstra, 2008).

There is, however, a compromise. The bioreactor technology most familiar to the industry can be used in combination with microcarriers to provide cells with an adherent environment to grow on in a suspension culture vessel (explored in more detail in Chapter 4). Microcarriers are small spherical coated or porous beads that provide cells with an environment to adhere to and on which to grow whilst capitalizing on the well-mixed environment of a bioreactor. This well-mixed environment provides optimized exposure to nutrients in the cell culture media and oxygen required for growth and also limits exposure to growth inhibitors produced through cell metabolism. Increasingly, plastic and composites are now being used to push yields higher and aid cell recovery, including the use of thermoresponsive polymers.

1.4.5.1 Stirred Tank Bioreactors

Stirred tank bioreactors are the most common bioreactors used in the biopharmaceutical industry and have been employed with some success in the production expansion of human cells for cell therapies in combination with microcarriers (Kirouac and Zandstra, 2008). Stirred tank bioreactors contain impellers of various designs that gently mix the culture, keeping the cells adhered to microcarriers in suspension, whilst also allowing nutrient and gas transfer. Without a significant step change in unsupported suspension culture (Kirouac and Zandstra, 2008), stirred tank bioreactors and microcarriers are currently the only plausible hope for mass production of cell therapy products (Simaria et al., 2014).

1.4.5.2 Rocking Platforms

In 1996, Vijay Singh invented the "Wave" bioreactor (Singh, 1999). These rocking platforms possess all of the benefits attributed to stirred tank bioreactors without the need for an impeller, thus reducing shear forces and the risk of mechanical breakdown. The more gentle approach to mixing is often more suited to cells that are sensitive to shear stresses, as human cells are. Though rocking platforms are limited in scale they are often used in the seed train (the pre-culture needed to reach large production scale in stirred suspension bioreactors using microcarriers) and for smaller-scale production for patient-specific products.

1.4.5.3 Perfusion Cell Culture

Perfusion cell culture is fast emerging in cell therapy processing. It allows cells to grow to higher densities in bioreactors by providing fresh medium and removing waste products continuously. Perfusion devices are attached to the bioreactor and provide low shear environments that can be used for media exchange and cell removal in continuous cell culture. Continuous cell culture is the process by which cell removal and media replacement take place continuously to extend the cell culture time and also productivity, that is, the cell number to medium ratio. This method has merits in efficiency and reducing the downtime of equipment and thus increasing the amount of cells manufactured per unit time.

Perfusion cell culture can also be used for washing and concentration of therapeutic cells. Media exchange from one formulation to another can occur in a perfusion reactor and cell concentration is also possible by reducing the ratio of inflow to outflow (Box 1.9). It is a low shear process that reduces the potential for contamination, as the culture broth is not exposed in a separate unit operation.

1.4.6 Differentiation

Differentiation is the process by which pluripotent stem cells are forced down a specific lineage to produce the cell of interest, in this

Box 1.9 Alternative tangential flow

The Alternative Tangential Flow (ATF) system from Refine Technology is a perfusion module that can be added to a stirred tank bioreactor. The module consists of diaphragm, which is pumped up and down, connected to a filter housing and then attached to a traditional bioreactor. Medium exchange can be easily performed using the ATF system, leading to higher cell density batches. The ATF device can be used with both cell and cell-microcarrier systems. When the ATF device is used with a system that uses microcarriers to grow cells, it is possible to perform further process steps in a single bioreactor (e.g. enzymatic detachment of cells from microcarrier beads) and thus reduce downstream processing. These steps are concentration, wash, perfusion, rapid medium exchange, trypsinization and cell-microcarrier separation (www.refinetech.com).

case the therapeutic cellular therapy product. Often the cell of interest would not proliferate or would do so very slowly and so the precursor cell to the type required is cultured to the quantity desired and then differentiated into the cell type of interest. Differentiation is a process step that is carried out by making additions to the culture media that force the pluripotent or multipotent stem cells to turn into the required cell product. Beyond this step, in the downstream processing, care must be taken to remove any undifferentiated cells from the therapeutic dose. Pagliuca et al. (2014) have demonstrated that differentiation is possible at scale. They were able to differentiate, in suspension, at a 500 mL scale, human pluripotent stem cells into functional pancreatic beta cells that comprised around 10% of the final cell population.

1.5 Downstream Processing of Cellular Therapies

1.5.1 Harvest, Washing and Concentration

After cell expansion, the next steps in cell therapy processing are to wash cell culture media from the cells, re-suspend in a non-proliferative media and concentrate into a volume more suited to further processing. The techniques traditionally used derive from the blood processing industry, but there is a desire to move towards the more scalable methods of traditional biopharmaceutical processing, particularly for allogeneic cell therapies. These techniques are discussed in detail in Chapter 5.

1.5.1.1 Centrifugation

Centrifugation is a unit operation that is traditionally used to separate cells from the supernatant and concentrate cells after cell culture. Centrifugation is used in traditional biopharmaceutical manufacture where the cells are no longer required and thus the processing step is designed to collect the protein present in the supernatant. Traditionally then, this step is used to remove the cells as opposed to preserving them.

Batch centrifugation is the most commonly used laboratory method. Cells can be separated from the supernatant using centrifugal forces in a batch approach. This has some shortcomings when used for cell therapy manufacture. Firstly, the removal of cells from the centrifuge bowl can case cell damage and, secondly, the batch centrifugation process is limited by scale.

Continuous centrifugation (described in Chapter 5) has been used with success in the processing of cell therapies. Orthogonal continuous centrifugation, where fluid flow is perpendicular to the centrifugal force, can be used to concentrate cells but is limited in its washing capability. Counterflow centrifugation, where fluid flows in the opposite direction to the centrifugal force, allows both cell washing and concentration. Counterflow continuous centrifugation has been developed and used with success in cell therapy manufacture; its scalable nature and use of a single-use fluid path provides a suitable solution for the gentle washing and concentration of human cells (Pattasseril et al., 2013).

1.5.1.2 Filtration

The use of normal flow filtration (where pressure and fluid flow is perpendicular to the filter) and tangential flow filtration (where fluid flow is along and parallel to the filter and a perpendicular pressure is applied) is routine in traditional biopharmaceutical processing. However, normal flow filtration is generally used for sterilization and does not handle large volumes of liquid well. Tangential flow filtration is used for separation processes in biopharmaceutical manufacture, but it employs high trans-membrane pressure that is not suitable for cell therapies, as it would compromise cell viability due to high shear forces. Industry has now developed tangential flow filtration devices with low shear effects through the application of low transmembrane pressures that are suitable for use in both concentrating and washing human cells (Pattasseril et al., 2013).

Filtration devices range from the most basic to fully automated and as they are mostly disposable they will integrate well into a cell therapy manufacturing process. The capital set-up cost for filtration devices is significantly less than for centrifugation. In addition, the fluid path remains enclosed, so risk of process failure is lower, and environmental background requirements are also lower than for centrifuges. Such potential advantages may lead to filtration devices becoming more prominent in cell therapy processing (Pattasseril et al., 2013).

1.5.2 Separation and Purification

After cell concentration and media exchange, the final steps in cell therapy processing are separation of the desired cells and purification to remove any unwanted residuals in the solution. For traditional biopharmaceuticals, a combination of filtration and chromatography steps are used. For cell therapies, density gradient centrifugation and/or magnetic separation are the preferred methods (Pattasseril et al., 2013). Though an outline is provided on current cell separation technologies here, this topic is covered in more depth in Chapter 5, where novel methods will also be discussed.

1.5.2.1 Centrifugation

Centrifugation is used to purify cellular products as well as in cell recovery. Centrifugation separates the bulk mixture based on its composition in terms of density and size. This process step is designed so that the fraction that contains the cells of choice is harvested for further processing. Density gradient separation has been made more accurate through operation in counter flow mode. Centrifuges have been used to process blood for some time, an example of this being the Terumo BCT's Cobe 2991 system that performs blood apheresis using continuous centrifugal technologies. The scale of this system is limited to less than 1 litre and it is best used for one to one cell therapies. In traditional biopharmaceutical production, centrifuges are widely used. However, they are focused on the removal of cells from the supernatant that contains the product. As the cells are a by-product in traditional biopharmaceutical manufacture there is little requirement to keep the cells intact, with the caveat that cell disruption can make subsequent downstream processing steps more difficult. The kSep from KBI is a low shear centrifuge system that has shown great promise in cell therapy processing Box 1.10). It is a continuous centrifuge with a disposable flow path that achieves 80% cell recovery whilst maintaining greater than 90% cell viability (Pattasseril et al., 2013) and is able to process 720 L/h (www.ksepsystems.com).

Box 1.10 Approaching the problem from two directions

The first cell therapies to be adopted used bone marrow stem cells for which a manufacturing approach is well established. In the UK, this manufacturing is carried out in hospitals and each batch is processed separately. Though processing blood for blood transfusion is centralized, each unit of blood that is donated is also processed separately. This means a raft of technologies have been developed to service blood and bone marrow processing, which is effectively a one to one therapy. These technologies have made their way into cell therapy processing due largely to hospitals and universities being the prime location for early stage cell therapy development (Brandenberger et al., 2011). Centrifugal techniques, such as Terumo BCTs Cobe 2991 (www.terumobct.com), have been used in cell therapy processing. The scale for which these technologies were developed however, means their application in cell therapy processing at scale is limited.

The biopharmaceutical industry is used to manufacture large batches of therapeutic proteins and technologies have been developed to complete the steps required for these one to many therapies. Though not entirely suited to the need to retain the cells as the therapeutic product, these unit operations and processes have been developed and used by industry to produce large batches under comparatively strict manufacturing regulations to those for ATMPs. Centrifugal technologies such as Pnuematics Unifuge (www.psangelus.com) and KBIs kSep (www.kbibiopharma.com) have been developed using the principles of the biopharmaceutical industry, whilst also dealing with the sensitivities around the cells being the final product.

Cell therapy bioprocessing is dealing with both of these challenges, as some products are one to one and some are one to many, depending on the therapy in question; however, all must comply with the regulations of drug manufacture.

1.5.2.2 Magnetic Separation

Magnetic separation is analogous to affinity chromatography steps in traditional biopharmaceutical manufacture in that it uses a property of the surface of the cell required to perform a highly accurate separation (Box 1.11). Identification of a surface antigen unique to the cell of choice is required. An antibody specific to this antigen is attached to a magnetic particle that binds to the cells of choice when they are exposed to each other. Thus the cells of choice are retained by a magnetic field and all waste product passes through the column.

1.6 Formulation, Fill and Finish of Cellular Therapies

Unlike traditional biopharmaceuticals, which are usually packaged for delivery to the patient as either a vial or pre-filled syringe for percutaneous injection or as a sterile bag for infusion, cell therapies are administered to patients in many different ways; they may be

Box 1.11 CliniMACs

Miltenyi Biotech was founded in 1989 by Stefan Miltenyi to commercialize the Magnetic-Activated Cell Sorting (MACS) technology he had invented (www.miltenyibiotech.com). The technology involves coupling a magnetic bead to a cell of interest using an antibody specific to that cell. They are then separated from other cells using a magnetic field. The CliniMacs system which automates this process was developed for cell selection processes, but has also found a market for cell separation in bioprocessing of cellular therapies. It has been approved by the FDA for CD34 selection in the treatment of acute myeloid leukemia. The MACs technology has been widely adopted by scientists and has been the subject of over 20,000 publications and has been used in over 40,000 clinical treatments. The CliniMACs has now been further developed and the CliniMACs Prodigy was released in 2013. The CliniMACs Prodigy is a GMP compliant cell processing device that completes both upstream and downstream processing steps in a closed, automated system. The culture chamber of this system only holds up to 400 mL and thus this system is more suitable for one to one therapies.

administered as a sheet of cells, an injection, intravenously or in a more complex operative procedure. This means the final form in which they are delivered to the clinic differs from one therapy to another (see Chapters 6 and 7). However, they all share similarities in the complexity of their administration. Formulation, Fill and Finish is an often neglected area of cell therapy development. It is key, however, to ensuring an efficacious product is delivered. Poor stability can result in business models that are not cost-effective, negating process efficiencies achieved earlier in the process. To ensure robust and reliable delivery of cells, an End to End Supply Chain mentality needs to be utilized, as exemplified by the Seamless Freight activities at the Cell and Gene Therapy Catapult (www.ct.catapult.org.uk).

1.6.1 Formulation

As a final step in processing, the cells should be formulated into a buffer that is suitable for storage. The most effective approach is to ensure that the buffer solution the cells are formulated into may also be suited to administration to humans; this approach is becoming more prominent in the industry. Where this cannot be the case, an additional step must be added in the clinic where a buffer exchange will be required.

1.6.2 Fill and Finish

The cell solution should now be added to the final storage and administration vessel. This could be a vial, a bag, a prefilled syringe or some novel storage device. Packaging live cells has challenges in terms of shelf life and the fragility of cells. The latter challenge requires that care be taken not to damage the cells through shear stresses on the final fill step in the cell suspension.

Many cell therapies are delivered to the patient in blood bags at low cell concentrations that are suitable for small lot sizes, but as lot size increases the use of vials and filling machines at much higher cell concentrations will be necessary (Brandenberger et al., 2011).

There is an increasing need to ensure process data is generated on hold times and temperatures, pre-fill mixing shear rates and fill speed shear, as these will have an impact on stability.

Shelf life of the cell suspension at this stage in the process will often be limited and thus this part of the supply chain must take place as quickly as possible. The volume and number of doses being processed will have an influence on what is feasible in terms of packaging times, but a general rule is "as fast as possible". There are no intermediate products that can be stored and this limits downstream processing choices and increases the need for real time fill. For some allogeneic cell therapies, large, high throughput filling machines whose overall capacity far exceeds the batch size are required in order to minimize the time for this operation. By way of an example, Provenge, which is a one to one cell therapy, has a shelf life of only 18 hours once packaged and 3 hours once removed from its insulated packaging. By comparison, if the same therapy was one to many and a batch were produced at scale in the bioreactor systems described previously (Simaria et al., 2014), then 4,708 vials would need to be filled; this process could take almost 8 hours using a system such as the Aseptic Technologies LX1 filling line (www.aseptictech.com), which can fill 600 vials per hour. This would significantly reduce the available shipping time for the product. However, a line such as Boschs FLC3000 (www.boschpackaging.com), which can fill 36,000 vials per hour, would take approximately 8 minutes and would have little effect on the available shipping time.

1.6.3 Preservation and Shipment

Over time the development of methods of preservation such as lyophilization have been key to the supply chain of traditional biopharmaceuticals. However, for cell therapies, successful preservation strategies are still very much in development. Cell storage solutions are covered in detail in Chapters 6 and 7.

For cell therapies whose product shelf life cannot be increased through preservation, the outbound supply chain is critical, as the therapeutic product must reach the patient as soon as possible. This is more likely to be the case for an autologous therapy in many instances (e.g. Provenge, discussed above); for an allogeneic cell therapy, it does not make commercial sense to develop a product that cannot be preserved. There are, however, cases where allogeneic cell therapies have minimal preservation. For example, Apligraf in its final form has a shelf life of 5 days (Czaja et al., 2006) and thus significant amounts of product are often disposed of, because they have exceeded their expiry date, increasing overall cost of goods.

Shipment of cell therapies that are not preserved is usually carried out at low temperature to delay metabolism and cell death (Chapter 7).

Transportation methods exist for biopharmaceutical therapies, blood and organs. However, good distribution practice dictates that the mode of transport must ensure that the product has not been exposed to elements that may compromise their quality and integrity (Coopman and Medcalf, 2014). Temperature monitoring and control has historically been the biggest challenge; however, Nikolaev et al. (2012) also demonstrated the negative effects of vibration on mesenchymal stem cells stored at 4°C.

Some cell therapies can be cryopreserved (see Chapter 6). That is, they are mixed with a buffer that protects the cells and can be stored at ultra low temperatures (−178°C). Controlled rate freezers are used to gradually cool the cells to this temperature. Traditionally this is achieved through the use of liquid nitrogen. More recently, novel approaches have reached the market, such as Asymptotes (www.asymtote.co.uk) VIAFreeze technology that is based on Stirling engine cooling. Once cryopreserved, cells for therapeutic use can be stored for significant periods but the outbound supply chain should be considered; for example, local cryogenic storage temperatures may be higher than in the manufacturing facility. If reformulation is not carried out in the clinic once the therapeutic dose comes out of cryopreservation, this must be done prior to shipping. Shipment periods should then be kept to a minimum.

Preservation of cell therapies not only reduces the risk in the outbound supply chain but also allows time for quality control test results to be returned and the product released by a Qualified Person with a complete dataset and reduced risk. Very short shelf lives mean that some key quality control data are often returned after release and administration to patients; this is less than ideal.

1.7 Administration of the Cell Therapy to a Patient

The final steps required to deliver a cell therapy to the patient have the potential to be the weakest link in the supply chain. A major challenge is consistency between sites during the process of preparing the cell therapy for administration to the patient. Ideally there should be no reformulation at the bedside. Even without reformulation there is a possibility of variability, for example, cryopreserved products require thawing and this may not be carried out reproducibly from one clinical site to another. However, automation to control and record cell thaw and low-risk preparation steps are being developed by companies in the field. A long-term aim for this final step in the supply chain must be automation to reduce variability and human error to a minimum.

1.8 Cell Therapy Manufacturing Facilities of the Future

Given the history of the emergence of the cell therapy industry, existing manufacturing facilities are largely designed to accommodate processes that are open and compromised by non-sterility. This means that the majority of manufacturing suites are small, high-grade clean rooms where processes are manually undertaken under controlled conditions. This route to manufacture is more suited to one to one therapies, where the production of each dose is a separate process and therefore should be self-contained. When considering this process route for one to many therapies, its labour intensity has to be markedly amplified to achieve the volumes of doses needed.

The ability to produce cell therapies on an individual or small batch basis in these high-grade clean rooms has been the basis of manufacture to date and has served the needs of early phase clinical trials adequately. However, as the industry matures, as there are more cell therapies in development and as a greater number reach later stage clinical trials, so there is a need to address the current *status quo* for manufacture.

Current processes are, in the main, open. That is, as the process medium moves from one unit operation to the next, it is exposed to the surrounding environment and possible contamination. In order to satisfy regulatory requirements, there is a clear need for high-grade clean rooms and highly trained personnel. If cell therapies are to be manufactured at scale there is, however, a pressing need to move away from a processing route that consists of these elements (high-grade clean environments, open processes and highly trained personnel) to a manufacturing route that reduces both risk and cost of goods, yet still produces therapies that are of the highest quality through closed processes and automation.

In order to move to this next stage, there are some key challenges. The first is that the production methods and scale of production for cell therapies varies widely, from one to one to one to many, from adherent to suspension or scaffolds, from orphan diseases to highly prevalent diseases and so on. The variety of cell therapy products means that a single, standard platform process is not feasible. However, it is possible to foresee a future where a suite of standardized platforms may be available or rather that a range of unit operations available at different scales may be plugged together to form a whole process. The second challenge is the ability to move towards closed processing. If it is possible to close a process so that it is not exposed to the external environment then lower-grade clean room suites can be used. This would then reduce the current high costs of such an environment and would also allow multiple processes to take place at the same time, in the same suite. However, this is challenging for a number of reasons (see Box 1.12).

There is also the challenge of scale. For large-scale production of one to one therapies, the required step is to scale-out the process (i.e. parallel execution of the same process at the same scale occurring at the same time). For one to many cell therapies, there is a need to

Box 1.12 Challenges in moving to closed processing

The current labour intensity of existing processes means that they are subject not only to high specialized labour costs but also subject to risk through operator intervention and variation. Removing the labour intensity of processes via automation and the development of new process technology will help mitigate this cost and risk:

- when dealing with a range of unit operation technologies, as each equipment manufacturer has its own unique inlet and outlet configurations. Fluid paths need to be compiled and shown to be validated.
- There are a lack of industry standards to enable connectivity; both physical with regards to tube interfaces, but also equipment connectivity with regards to data integrity.
- Introduction of a process changes may lead to the need to complete product comparability studies to demonstrate no adverse affect on product.

scale-up the process to achieve production at large scales to satisfy the market demand.

As previously mentioned, manufacturing processes for cell therapy have often been born from blood processing suites in hospitals where the procedure happens close to the patient. This route of manufacture limits the supply chain demands, but will not be suitable or possible for all cell therapies. For one to many cell therapies, it is essential from a cost of goods perspective to follow the processing route of traditional biologics that are manufactured at a central site and then shipped to the patient. This is limited somewhat by the current shelf life of cell therapies, though investment is being made in developing longer shelf life and more stable products (see Chapter 6). It is doubtful that one supply chain will be developed for all cell therapies due to the variation in cell source, disease prevalence and the product's shelf life; thus various solutions have been proposed such as distributed manufacture (local, smaller, manufacturing facilities) or wholly closed automated processes that manufacture what is needed for each patient at the bedside or in the clinic. The latter approach challenges the current structure of the drug manufacturing industry, where each facility that manufactures any sort of drug must be licensed by the regulatory body and each batch of therapy must be released by a qualified person. One approach would be to use remote monitoring by qualified persons and thus real time release. Perhaps in the future the manufacturing devices that sit in the clinic may be classed and regulated as medical devices.

In order to achieve wholly closed automated processes, investment is needed in scale-down, automation, non-invasive sensing, data integration and management Box 1.13). This must be underpinned by

Box 1.13 Invetech

The concept of wholly closed and automated processes has been addressed by Invetech (www.invetech.au). Invetech have developed a closed processing platform for Argos therapeutic cancer immunotherapy which can be operated in a lower-grade clean environment. The process is wholly automated and requires only a small tumour sample from the patient. The therapy is now in Phase 3 clinical trials and demonstrates that early investment in process development and plans for routine manufacture has led to a suitable scalable solution that could indeed operate in the clinic or close to the patient.

predictive manufacturing so that quality is ensured throughout the process and not tested in and at the end by quality control requiring sign off by a qualified person, as is the current model.

1.8.1 Factory of the Future Requirements

Facilities of the future must be adaptable to markedly different product types and scales. Process development must be focused on efficient downstream processes as well as optimizing upstream growth, seamless fill finish, chill, freeze, and possible inactivation. They must also deliver frozen or chilled dispatch and straightforward logistics. The advantages are cheaper air handling equipment that need deliver only a lower level of air grade and much more reliable replication of manufacture, allowing focused delivery of product and the benefit of de-risked future process expansion.

As the industry is still young and there are as yet very few cell therapies on the market, it is impossible to say which manufacturing routes will become dominant. However, there is a commonly held view in the industry that open, labour-intensive processing in expensive clean rooms will not adequately service the industry, as more and more cell therapies come through the development pipeline and reach the marketplace.

1.9 Conclusion

The curative potential of cell therapies is showing great promise and thus the cell therapy industry continues to grow as new products emerge on the market and more products enter late phase clinical trials.

The challenges presented by one to one and one to many cell therapies are different; they require different supply chain approaches while adhering to the same manufacturing regulations. Though there are parallels to be drawn with the manufacture of blood products and traditional biopharmaceuticals, cell therapies are the most complex products to manufacture, formulate and administer in the history of therapeutics manufacturing.

In order for the cell therapy industry to realize its ambitions of providing curative therapies, the industry must address the pressing challenges of closed processing through interoperability of machines that carry out unit operations and develop robust manufacture at scale to ensure therapies are cost-effective and widely available.

References

Brandenberger, R., Burger, S., Campbell, A. *et al.* (2011) Cell therapy bioprocessing. *BioProcess Int.*, **9**, 30–7.

Carroll, J. 2013. Organogenesis preps "heart breaking" cuts as Medicare slashes reimbursement [Online]. Available at: http://www.fierce-biotech.-com/-story/-organogenesis-preps-heart-breaking-cuts-medicare-slashes-reimbursement/2013-12-04

Connolly, A. (2002a) Novartis squares off over Organogenesis' Apligraf. *Boston Business Journal*. http://www.bizjournals.com/boston/stories/2002/10/07/story8.html

Connolly, A. (2002b) Product matters: Organogenesis gets "Goliath" to blink. *Boston Business Journal*. http://www.bizjournals.com/boston/stories/2002/12/02/newscolumn3.html

Coopman, K. and Medcalf, N. 2014. From production to patient: Challenges and approaches for delivering cell therapies. *In: StemBook*, ed., The Stem Cell Research Community, StemBook, doi/ 10.3824/stembook.1.97.1.

Czaja, W.K., Young, D.J., Kawecki, M. and Brown, R.M. (2006) The future prospects of microbial cellulose in biomedical applications. *Biomacromolecules*, **8**, 1–12.

De Bie, C. (2007) Genzyme: 15 years of cell and gene therapy research. *Regen. Med.*, **2**, 95–97.

FDA. 2014. Vaccines, Blood and Biologics – Marketed Products [Online]. Available at: http://www.fda.gov/BiologicsBloodVaccines/CellularGeneTherapyProducts/ApprovedProducts/

Foley, L., Callens, S. and Osborne, A. (2012) Institutional profile: The ATMP Manufacturing Community (AMC). *Regenerative Medicine*, **7**, 285–9.

Foley, L. and Whitaker, M. (2012) Concise review: Cell therapies: The route to widespread adoption. *Stem Cells Translational Medicine*, **1**, 438–47.

GenEngNews 2014. "Shire Sells Money-Losing Dermagraft to Organogenesis".

Hambor, J.E. (2012) Bioreactor design and bioprocess controls for industrialized cell processing. *BioProcess Int.*, **10**, 22–33.

Hipp, J. and Atala, A. (2008) Sources of stem cells for regenerative medicine. *Stem Cell Reviews*, **4**, 3–11.

Jones, A. (2011) Shire buys Advanced BioHealing for $750 m. *Financial Times*. https://www.ft.com/content/a494712e-811e-11e0-9360-00144feabdc0

Kanczler, J.M., Ginty, P.J., Barry, J.J. *et al.* (2008) The effect of mesenchymal populations and vascular endothelial growth factor delivered from biodegradable polymer scaffolds on bone formation. *Biomaterials*, **29**, 1892–900.

Kemp, P. (2006) History of regenerative medicine: Looking backwards to move forwards. *Regen. Med.*, **1**, 653–69.

Kirouac, D.C. and Zandstra, P.W. (2008) The systematic production of cells for cell therapies. *Cell Stem Cell*, **3**, 369–81.

Li, M.D., Atkins, H. and Bubela, T. (2014) The global landscape of stem cell clinical trials. *Regenerative Medicine*, **9**, 27–39.

Lysaght, M.J. and Hazlehurst, A.L. (2004) Tissue engineering: The end of the beginning. *Tissue Engineering*, **10**, 309–20.

Mason, C. and Dunnill, P. (2009) Quantities of cells used for regenerative medicine and some implications for clinicians and bioprocessors. *Regenerative Medicine*, **4**, 153–7.

Mount, N. 2014. Clinical Trials Database Commentry.

Nikolaev, N., Liu, Y., Hussein, H. and Williams, D.J. 2012. The sensitivity of human mesenchymal stem cells to vibration and cold storage conditions representative of cold transportation. *Journal of The Royal Society Interface*, **9**, 2503–15.

Oancea, G., Wagner, B. and Henschler, R. (2012) Cell therapy regulations from a European perspective, in *Regenerative Therapy Using Blood-derived Stem Cells*, Springer Science and Business Media BV Dordrecht, Netherlands, pp. 191–204.

Pagliuca, F.W., Millman, J.R., Gürtler, M. *et al.* (2014) Generation of functional human pancreatic β cells *in vitro*. *Cell*, **159**, 428–39.

Pattasseril, J., Varadaraju, H., Lock, L. and Rowley, J. (2013) Downstream technology landscape for large-scale therapeutic cell processing. *BioProcess International*, **11**, 46–52.

Pearce, K.F., Hildebrandt, M., Greinix, H. *et al.* (2014) Regulation of advanced therapy medicinal products in Europe and the role of academia. *Cytotherapy*, **16**, 289–97.

Reporter, B. 2014. Shire takes 650m hit in regenerative skingraft divestment [Online]. Available at: http://www.biopharma-reporter.com/-Upstream-Processing/Shire-takes-650m-hit-in-regenerative-skingraft-divestment

Simaria, A.S., Hassan, S., Varadaraju, H. *et al.* (2014) Allogeneic cell therapy bioprocess economics and optimization: Single-use cell expansion technologies. *Biotechnology and Bioengineering*, **111**, 69–83.

Singh, V. (1999) Disposable bioreactor for cell culture using wave-induced agitation. *Cytotechnology*, **30**, 149–58.

Smith, D.M. (2014) Practicalities to translation from the clinic to the market, in *Chemical Biology in Regenerative Medicine: Bridging Stem Cells and Future Therapies*, John Wiley & Sons, Chichester, UK, pp. 203–15.

Thomas, E.D., Storb, R., Clift, R.A. *et al.* (1975) Bone-marrow transplantation. *New England Journal of Medicine*, **292**, 832–43.

Tozer, D. (2011) Interview: The story of Advanced BioHealing: Commercializing bioengineered tissue products. *Regenerative Medicine*, **6**, 157–61.

Vanlauwe, J., Jelic, M., Limbourg, M. and Luyten, F. 2014. Characterized chondrocyte implantation challenges current paradigms for the treatment of symptomatic joint surface lesions. *In: Developing Insights in Cartilage Repair*, Springer Science and Business Media BV, Dordrecht, Netherlands. Ebook 9781447153856.

Wei Teng, C., Foley, L., O'Neill, P. and Hicks, C. (2014) An analysis of supply chain strategies in the regenerative medicine industry – Implications for future development. *International Journal of Production Economics*, **149**, 211–25.

Willerth, S.M. and Sakiyama-Elbert, S.E. 2008. Combining stem cells and biomaterial scaffolds for constructing tissues and cell delivery. *In: StemBook*. Harvard Stem Cell Institute, Cambridge, MA. Available at: http://www.ncbi.nlm.nih.gov/books/NBK27050/ doi: 10.3824/stembook.1.1.1

2

Structured Methodology for Process Development in Scalable Stirred Tank Bioreactors Platforms

Huaqing Wang[1], Daniel Kehoe[2], Julie Murrell[2] and Donghui Jing[3]

[1] *Department of Medical Oncology, Nankai University Cancer Hospital, Tianjin Union Medical Center, Tianjin, P.R. China*
[2] *Stem Cell Bioprocessing Group, Process Solutions, EMD Millipore Corporation, Bedford, USA*
[3] *HaoLing Cell Technologies Corporation, TEDA, Tianjin, P.R.China*

2.1 Introduction

Stem cell-based cell therapy holds great promise to achieve the next generation of medical breakthroughs. According to ClinicalTrials gov, 126 reported trials using multipotent progenitor cells, such as human Mesenchymal Stem Cells (hMSCs), are underway for a variety of indications, such as graft-versus-host disease (GVHD), acute myocardial infarction, Crohn's disease and multiple sclerosis. The intensive usage of hMSCs for allogeneic or autologous therapies is mainly due to their capability to differentiate into various cell types as well as their immunologic properties. More importantly, allogeneic therapies have the advantages of the possibility for centralized manufacturing, storage and distribution, which is similar to classical therapeutic agents such as small molecules or biologics.

However, a major hurdle in the clinical application of cell therapy is the high cost of the manufacturing process (Jon Rowley, 2012). Moreover, the traditional two-dimensional (2D) culture of adherent stem cells is not only labor intensive but also lacks constant monitoring of cells during the production process. Therefore, 2D cultures can only be regarded as sub-optimal for the manufacturing of clinical grade stem cells. To overcome these deficiencies the use of stirred tank

Bioprocessing for Cell Based Therapies, First Edition. Edited by Che J. Connon.

bioreactors combined with microcarriers provides an attractive alternative. Scale-up in bioreactors with microcarriers is not a new challenge, and has been utilized in industry for decades for cellular synthesis products, such as therapeutic proteins and vaccines. For example, a few reports have been published to describe the cultivation of hMSCs on various microcarriers (Eibes et al., 2010; Hewitt et al., 2011; Schop et al., 2008). However, most of these studies were done in lab-scale small bioreactors, for example spinner flasks. Until recently, there were reports showing that several groups were able to scale up of the process in liter-sized bioreactors (Dos Santos et al., 2014; Jing, 2013; Rafiq et al., 2013a, b). Here, we discuss primarily the use of stirred tank bioreactor platforms for hMSCs cultivation, because these vessels are extensively used in biotechnology and pharmaceutical companies.

In this chapter, we will focus on:

- understanding the engineering of the stirred tank bioreactors, such as mixing phenomena, oxygen transfer rate (k_La) and heat transfer;
- understanding the biology of the cells, such as cell types, assays for comparability, nutrients, metabolism and sensitivity of cells to culture additives; and
- process development of adherent cells, mainly hMSCs, in stirred tank bioreactors.

Finally, we will discuss the challenges lying ahead that need to be addressed.

2.2 Understanding the Engineering of the Stirred Tank Bioreactors

Several requirements must be met to provide the proper environment for the culture of cells. This section focuses on some of those requirements including agitation, oxygenation and heat transfer. Many of the commercially available stirred tank bioreactors were designed for single cell suspension cultures and there are some nuances with regard to culturing of cells for cell therapy that are discussed throughout this section. The engineering of a culture surface for cells that require surface attachment is also examined. The aim is to provide a better understanding of the specific needs of the cell therapy bioprocess in relation to bioreactor design or selection.

2.2.1 Mixing Phenomena in Stirred Tank Bioreactors

Impeller design and bioreactor geometry are important factors that determine the mixing properties of a stirred tank bioreactor. Mixing time, power provided per unit volume, microcarrier settling and eddy length are some of the factors to consider when selecting an agitation rate, changing the agitation rate to account for an increase in the working volume of a bioreactor, or when scaling between differently sizes or types of stirred tank bioreactors.

There are several different impeller types that are used in stirred tank bioreactors, including rushton, pitched blade, elephant ear or 3-blade segmented, and marine impellers (Figure 2.1). The rushton impeller is a flat blade (paddle) impeller that pushes fluid in the radial direction, which is generally run in the turbulent flow range to also create some fluid flow in the axial direction. The iconic example of a stirred tank bioreactor that uses rushton impellers is the spinner flask from Corning. Pitched blade impellers are flat blade impellers that are angled (typically from 5 to 45 degrees) to provide fluid flow in both the axial and radial directions. Examples of single-use bioreactors that utilize pitched blade impellers include Mobius® CellReady (50L and 200L) bioreactors from EMD Millipore, HyPerforma™ (50L and 250L) bioreactors from HyClone, Xcellerex XDR (50L, 200L, 500L, 1,000L and 2,000L) bioreactors from GE Healthcare Life Sciences, and CelliGen® BLU (5L, 14L and 50L) from Eppendorf. Elephant Ear or 3-blade segmented impellers can provide more surface area compared to the same sized pitch blade impellers, which may provide shorter mixing times.

Bioreactors and fermenters that utilize elephant ear or 3-blade segmented impellers include products from ABEC and the BIOSTAT® STR family of bioreactors (250 mL to 2000 L) from Sartorius (although

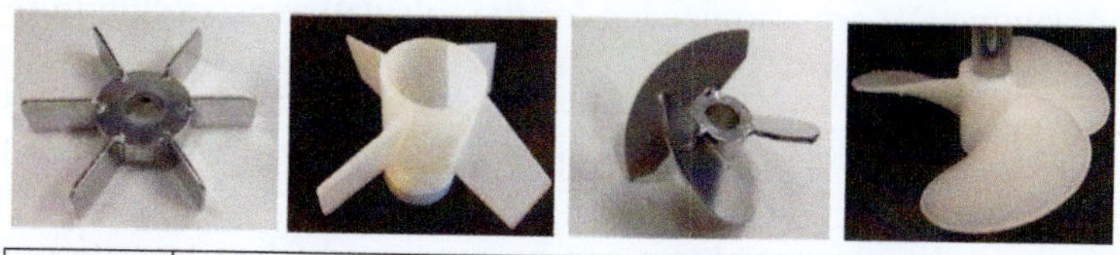

Name	Rushton	Pitched Blade	Elephant Ear/ 3-Blade Segmented	Marine
Fluid Flow	Radial	Radial and Axial	Axial	Axial

Figure 2.1 Different impeller types used in stirred tank bioreactors (courtesy of EMD Millipore Corporation).

the Sartorius bioreactors generally employ 2 impellers; both a 6-blade rushton and 3-blade impeller or two 3-blade impellers). Marine impellers provide fluid flow primarily in the axial direction. The Mobius® CellReady 3L bioreactor from EMD Millipore makes use of these impellers. While this is not an exhaustive list of all the different types of impellers, a large portion of the single-use stirred tank bioreactors currently utilizes these types and further discussion of power input and mixing will be limited to these. A more complete list of all the current single-use bioreactor systems offered commercially, including 2D bioreactors, was recent published by Loffelholz et al. (2013).

The first mixing parameter to consider is mixing time. This can help to determine how quickly a bolus addition of liquid can be distributed throughout a bioreactor or help to compare the impellers' ability to mix in a broad sense between different scales or bioreactor types. This information can usually be found on the bioreactor manufacturer's website, but it may not necessarily apply if the agitation rates needed to grow cell therapy products are lower than the rates used to determine the mixing times.

Another important process engineering parameter to aid with scaling is power input per volume. Power input is the amount of power that can be applied to the impeller to create rotational speed. Generally when more volume is added or if different scales of bioreactors are being compared, it can be helpful to make the comparisons keeping the power input per volume equivalent rather than maintaining a constant agitation rate. To calculate the power input per volume (P/V) a power number (N_P), which is a property of the impeller, is required. This is generally supplied by the manufacturer. In general, rushton impellers have higher power numbers than axial flowing impellers. The diameter of the impeller (d), fluid density (ρ), rotational speed (n) and volume (V) are also needed for the calculation according to the following equation:

$$P/V = N_p n^3 \rho d^5 / V \tag{2.1}$$

The power input per volume can be conserved when transitioning between different vessel scales. Even if the different vessels have different impeller types, the fluid within both bioreactors can experience the same amount of power per volume if the power numbers are known and the agitation rates are determined based on the power input equation. The agitation required after increasing volume (at time of a feed) of a bioreactor can also be determined by conserving power input per volume. When the volume of a bioreactor is increased from V_1 to V_2, then agitation can be increased from n_1 to n_2 based on the following equation:

$$n_2 = \left(n_1^{\ 3} V_2 / V_1\right)^{1/3} \tag{2.2}$$

Since microcarriers are generally used for the expansion of adherent cell therapy products in stirred tank bioreactors, there is another important parameter to consider when choosing an agitation rate for scaling: microcarrier settling. Increasing agitation rate based solely on power input per volume could lead to microcarriers falling out of suspension, thus it is also recommended to conduct mixing studies to determine the minimum agitation rate to maintain the microcarriers in a just suspended state or n_{js} (Figure 2.2). Uniform suspension of microcarriers may not be required for a successful culture, but generally when microcarriers pool on the bottom of the reactor there is less surface area available in the bulk of the bioreactor for the cells to grow and there is an increased chance of having nutrient limitations where the cells are pooling and thus agitation rate below the n_{js} should be avoided. Ideally the agitation rate of the bioreactor is held slightly above n_{js} which can be experimentally derived though mixing studies. As cells grow, the density of microcarriers may increase, so observation of the bioreactor throughout a culture may be required to avoid pooling of the microcarrier on the bottom of the vessel.

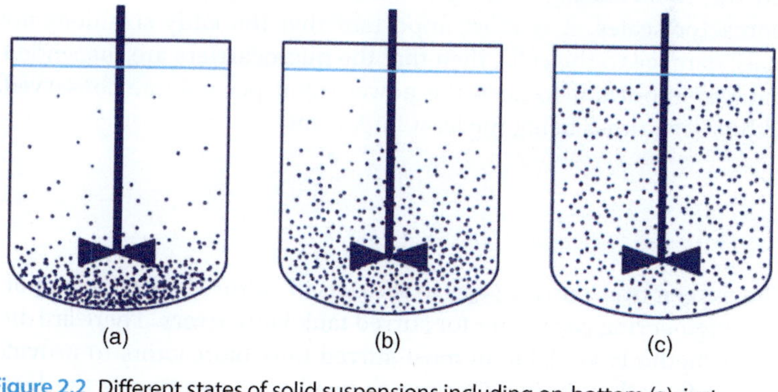

Figure 2.2 Different states of solid suspensions including on-bottom (a), just suspended (b) and uniform suspension (c) (courtesy of EMD Millipore Corporation).

Impellers that provide axial fluid flow are better equipped to prevent settling when compared to impellers that move fluid in the radial direction. The direction of axial flow can also change the mixing properties of the stirred tank bioreactor, but it is usually recommended to provide upward axial mixing (scoping) for microcarrier-based cultures, so the microcarriers are not forced down into the base of the bioreactor, which could be a cause of cell or micarocarrier damage.

The Kolmogorov eddy length should also be considered when choosing the agitation rate of a stirred tank bioreactor when growing cells on microcarriers. When the eddy length (η) is similar to the diameter of the microcarriers, then the fluid perturbations can cause damage to cells (Kehoe et al., 2010). These eddy sizes can be calculated based on the impeller properties and volume in the bioreactor according to the following equation, where ν is the kinematic viscosity:

$$\eta = \left(\nu^3 / \left(N_p d^5 n^3 / V \right)\right)^{1/4} \tag{2.3}$$

Because the eddy length decreases with agitation rate, it may be necessary to increase the agitation rate to ensure that the eddy length is below the microcarrier diameter range to avoid damage to the cells from eddies. The eddy length also increases with volume, which is another reason to increase the agitation when increasing the working volume.

Of the considerations to take into account when changing agitation rate due to increasing working volumes, changing bioreactor types or bioreactor scales, it is most important that the eddy size does not cause damage to the cells, then that the microcarriers are suspended, and less important to keep the power input per volume conserved, with mixing times being the least important.

2.2.2 Understanding Oxygen Transfer Rate (k_La) with Different Sparging Methodologies

The oxygen transfer rate (k_La) to the cells in culture is another important engineering parameter for stirred tank bioreactors. There are different methods available in most stirred tank bioreactors to provide oxygen for cell cultures. Overlay or headspace aeration is the least intrusive option. Based on the spinner flask model that is cultured in an incubator, overlay oxygen transfer can provide oxygen up to moderate cell concentrations (perhaps around 500,000 cells/mL); however, this can vary greatly depending on the oxygen uptake rate (OUR) of the cells and the k_La of the system. For higher density cultures, sparging may need to be implemented. Most single-use stirred tank bioreactors on the market provide both open tube and microsparging options. Microsparging is generally more effective for oxygen transfer, while an open tube is generally used in traditional bioprocessing to strip carbon dioxide.

While the oxygen transfer rate (k_La) in generally available by the manufacturer of a bioreactor, these values can vary greatly with the presence of serum, pluronic acid, and/or microcarriers in the culture media. Determination of the actual oxygen transfer rate for a specific system is recommended if high cell densities are desired for cell therapy cultures which utilize some of these additives. Once the oxygen transfer rate is known, the maximum cell concentration (X_m) can be calculated from the equation below, where C^* is the saturated dissolved concentration entering, C_L is the actual dissolved oxygen concentration in the vessel and q_{O2} is the specific uptake rate of oxygen:

$$X_m = k_La\left(C^* - C_L\right)/\,q_{O2} \tag{2.4}$$

The specific uptake rate of oxygen can generally be found in the literature, for example the q_{O2} for human mesenchymal stem cells has been reported to be 98 fmol/cell/h (Pattappa et al., 2011). Some k_La studies with microcarriers have been performed recently with the 3L Mobius® CellReady bioreactor to determine different ranges of k_La that can be mapped out with different oxygenation strategies (overlay, open tube and microsparging) at various agitation rates and gas flow rates. These k_La values were then calculated to determine the maximum cells (X_m) with either air ($C^*_{Air} - C_L$ = 0.25 mmol/L) or oxygen ($C^*_{O2} - C_L$ = 1 mmol/L) as the feed gas. The specific oxygen uptake rate for hMSCs of 98 fmol/cell/h was assumed for the calculations. The results from the calculations demonstrate that air overlay at 0.01 volumes of gas per volume of liquid per minute (vvm) or 10 ml/min is capable of supporting 250 million cells in a liter of media (Table 2.1). The maximum number of cells can be increased to 2 billion cells if the agitation is increased and the gas changed to oxygen. For typical hMSCs cultures, the maximum overlay aeration could theoretically

Table 2.1 k_La studies with microcarriers done in Mobius® CellReady 3L bioreactors.

	Agitation	Flow rate	kLa	Air		Oxygen	
Sparger Type	(rpm)	(vvm)	(h^{-1})	OTR (mol/ (L * h)	Maximum cells (millions)	OTR (mol/ (L * h)	Maximum cells (millions)
Overlay	50	0.01	0.1	0.000025	250	0.0001	1,000
Overlay	100	0.01	0.2	0.00005	500	0.0002	2,000
Open tube	100	0.01	0.5	0.00013	1,300	0.0005	5,100
Open tube	100	0.13	2.5	0.00063	6,400	0.0025	26,000
Microsparge	100	0.01	5.0	0.0013	13,000	0.005	51,000
Microsparge	100	0.10	20.0	0.005	51,000	0.02	200,000

result in the cell density of over 350,000 cells per cm^2, which is greater than what is generally observed in confluent 2D cultures.

2.2.3 Heat Transfer in STB (Minimum Volume, Sensor/Sensing Control)

Heat transfer is another engineering parameter that can vary with different designs of stirred tank bioreactors, but is very important to understand so that temperature can be accurately controlled during the expansion of cells.

Stirred tank bioreactors with less than 50 L working volume are generally heated with electric heating pads or blankets. Examples of these include the Mobius® CellReady 3L from EMD Millipore, UniVessel® 2L from Sartorius and CelliGen® BLU 5L, 14L and 40L from Eppendorf, and Xcellerex™ XDR-10 from GE. An advantage for these blankets is that they can change the temperature rather quickly, but they do not provide a large amount of insulation for the bioreactor. Most stirred tank bioreactors that are 50 L or larger use a temperature controlling unit to regulate the temperature of the water in the jacket to maintain the bioreactor temperature. The several gallons of water surrounding the bioreactor provide a great deal of insulation between the cell culture and the environment. Examples of these bioreactors include the Mobius® CellReady 50L and 200L from

EMD Millipore; BIOSTAT® STR50L, 200L, 500L, 1000L, and 2000L from Sartorious; Xcellerex™ XDR 50L, 200L, 500L, 1000L, and 2000L from GE; and HyPerforma™ Single-Use Bioreactor 50L, 250L from ThermoFisher.

There is a minimum working volume at which both electric blanket and water jacked bioreactors can no longer heat the cell cultures effectively or a minimum volume at which the culture cannot be agitated properly. Depending on the harvest strategy, it is important to consider the impact of this restriction on the ability to heat and stir during harvest activities, when small volumes of enzyme may be needed to dissociate cells from microcarriers when designing or choosing a bioreactor.

While the heat transfer capabilities of the blanket or jacket are designed to heat the bioreactor and maintain temperature, sometimes more complicated feed strategies or growth platforms may overwhelm these capabilities. Some perfusion devices require large volumes of media to flow in the alternating direction of a filter, which may decrease the temperature significantly and cause culture efficiency difficulties. Also some perfusion process steps use steady feeds or larger feeds during fluid exchange that may require preheating or higher demands from the heating system.

2.2.4 How to Choose a Microcarrier for Adherent Cells (hMSCs)

There is a wide variety of commercially available microcarriers that can support adherent cell therapy products in stirred tank bioreactors. Many of these microcarriers are shown in Table 2.2. The most important engineering parameter from a bioreactor design perspective of these microcarriers is the density. Some of the microcarriers are sold in different sizes, which have different densities to choose from. From a mixing point of view, the closer the density is to 1.00 g/mL, the easier it is to maintain suspension of the microcarriers and the lower the value of the just suspended agitation rate (n_{js}). The lower the n_{js} value, the larger the operational window of agitation rates can be. Also, since this value needs to be determined experimentally, n_{js} may be difficult to obtain for the larger scales when the microcarrier screening process begins. One advantage of heavier microcarriers is a faster settling rate for harvest steps, which could shorten the process during any settling steps that may be required in the harvest operations.

Table 2.2 Microcarrier property information.

No.	Manufacture	Name	Backbone	Coating	Charge	Surface Area (cm^2/gm)	Density (g/mL)
1	Pall	FactIII	Polystyrene	Gelatin	Cationic	360	1.02–1.03
2	Pall	Collagen	Polystyrene	Gelatin	None	360	1.02–1.03
3	Pall	Hillex® II	Polystyrene	Trimethylammonium	Cationic	515	1.09–1.15
4	Pall	Glass	High Silica Glass	High Silica Glass	None	360	1.02–1.03
5	Pall	Plastic	Polystyrene	None	None	360	1.02–1.03
6	Pall	Plastic Plus	Polystyrene	None	Cationic	360	1.02– 1.03
7	Pall	ProNectin	Polystyrene	recomb Fibronectin	None	360	1.02–1.03
8	GE	Cytodex®1	Dextran	None	Cationic	4,400	1.03
9	GE	Cytodex®3	Dextran	Collagen	None	2,700	1.04
10	Sigma	Cultuspher™ G	Gelatin	None	None	1,634	1.02–1.04
11	Sigma	Cultispher™ S	Gelatin	None	None	2,042	1.02–1.04
12	Corning	Enhanced Attachment	Polystyrene	Infused O^2		360	1.02–1.03
13	Corning	Synthemax® II	Polystyrene	Sythemax II	None	360	1.02–1.03

Of the microcarriers listed in Table 2.2, those manufactured by GE and Sigma are porous, while the ones listed from Pall and Corning are non-porous and on a backbone of polystyrene. For vaccine applications, porous microcarriers allow protection for the cells, since they can become embedded within the microcarriers. For cell therapy applications where the cells need to be easily released during harvest, a non-porous microcarrier may be the more suitable option. Although some groups have published data showing high attachment and growth of hMSCs on porous microcarriers, the cell recovery is not shown (Frauenschuh et al., 2007; Schop et al., 2010). A comparison of the harvested cell recovery from different microcarriers showed higher recovery from the polystyrene microcarriers compared to porous microcarriers (Kehoe, 2012).

Another consideration for cell therapy applications is the use of animal materials in the process and at least one group has moved away from the gelatin-based cultuspher™ microcarrier in favor of a polystyrene microcarrier (Santos et al., 2011).

In conclusion, there are several requirements of stirred tank bioreactors to provide a supportive culture of cells. This section discussed agitation, oxygenation, heat transfer and microcarrier properties. Many of the commercially available stirred tank bioreactors are designed for single-cell suspension cultures and some nuances regarding culturing of cells for cell therapy were discussed. The aim was to provide a better understanding of the specific needs of cell therapy bioprocessing in relation to bioreactor design or selection.

2.3 Understanding the Biology of the Cells in Stirred Tank Bioreactors (STB)

2.3.1 Cell Types (Adherent and Suspension Cells)

There are generally two categories of cells that are used in research and clinical settings. Those that grow while adhered to a surface and others that do not require adhesion to a surface for growth support (suspension). In addition, cells can either be the production machine (such as Chinese Hamster Ovary cells, CHO or Vero cells) or they can be the desired product, as is the case for cell therapy. Suspension cells were a natural first cell type to be investigated with the stirred tank bioreactor platform, as they can be introduced simply into a medium and mixed via an impeller. Adherent cells either require adaption to suspension culture, as was performed for CHO cells, a current workhorse of therapeutic recombinant protein production, or require the inclusion of a substrate, as is typically done for vaccine production. A variety of substrates are available to support cell growth; however, only a subset of these materials are appropriate for suspension culture and for recovery of the cells for therapeutic use. As recombinant protein and vaccine production has been thoroughly reviewed (Hess et al., 2012; Jostock and Knopf, 2012), we will focus on regenerative medicine applications for the remainder of this section. One special case for expansion in stirred tank bioreactors is cell aggregates. Some pluripotent stem cells (such as embryonic stem cells [ESCs] or induced pluripotent stem cells [iPSCs]) may grow well in an aggregate configuration, where they require attachment to other cells and these aggregates can be suspended through agitation to maintain sufficient nutrient and gas exchange.

2.3.2 Assays for Comparability, Nutrients, Senescence and Doubling Dime

The expansion method has the potential to affect the basic functions of the cells. Thus, it is important to fully characterize the cells. A collection of analytics to assess the process and the product is required to fully monitor the cells. The stirred tank bioreactor platform enables routine, safe sampling and monitoring. Process monitoring targets include physiochemical, nutrients, metabolites and essential amino acids. Key physiochemical variables that confirm the performance of the process include pH, dissolved oxygen, dissolved carbon dioxide, agitation rate and temperature. Nutrient uptake rates can be calculated for key nutrients including glucose and glutamine. And the corollary, metabolite production rates, can be calculated for compounds such as lactic acid and ammonia. As cell therapy bioprocessing matures, understanding the utilization of essential amino acids can contribute to the optimization of the culture, media and feed strategies.

Product monitoring targets include cell surface markers for identity and purity, cell growth, differentiation potential, senescence and potency assays. Flow cytometry is the workhorse for assessing the presence of cell surface markers that can be used to confirm identity (Dominici et al., 2006). Cell numbers can be captured through several mechanisms including total cell number, viable cell number and cell coverage of the microcarriers. When cell numbers are recorded on a routine basis, doubling times, fold increase and exponential growth rate can be calculated and identified. Differentiation potential, senescence and potency assays can all be considered as functional assays. Since the mechanism of action is not completely understood or identified for cell therapies, a collection of assays is used to provide the most complete picture possible. Differentiation potential can be assessed through traditional assays or more recently, through the use of immunoassay or nucleic acid assays (Deskins et al., 2013). Senescence is easily measured across a broad range of cells using a readout for beta-galactosidase (Goldstein, 1990). And, where they exist, potency assays should be included in a monitoring program to confirm the likely performance of the cell product.

In addition to the requirement of a substrate for growth of adherent cells in a stirred tank bioreactor, it is recommended to include antifoam and shear protectant in the growth media. Antifoam-C maintains a culture without foam, particularly important if sparging of gases is performed. The Pluronic® F-68 surfactant is a shear protectant commonly used in stirred tank bioreactors. While these additives are routinely used in CHO and Vero culture, it is important to determine the sensitivity of the cells to the two compounds.

2.4 Process Development of Adherent Cells in STB Platforms

2.4.1 Standard Comparison to Cell Factory

To date, large-scale production of bone marrow derived-hMSCs is typically performed using traditional 2D tissue culture vessels, such as 10-layer or 40-layer Nunc Cell Factories and Corning CellStack® plates. At EMD Millipore, we developed an internal protocol to expand hMSCs in 10-layer vessels for 2 weeks. Under the optimal condition, each 10-layer CellStack® plate produces approximately 200–400 million cells, depending on different donors (Jing et al., 2013a, b). A method to expand adipose tissue-derived stromal cells (ASCs) on multiplayer CellStack® plates under clinical Good Manufacturing Practices (cGMP) requirements was also reported (Bourin et al., 2011). However, taking into account the number of cells required for analytical assays, one vessel of

10-layer CellStack® plates might not generate one dose, assuming 250 million cells/dose. Overall, the application of 10-layer or 40-layer vessels can only enable lot sizes of 10–50 billion cells (Jon Rowley, 2012). These lot sizes may meet the needs of some autologous therapies or the clinical development stage for allogeneic therapies. More importantly, it is well believed that this method of manufacturing will not be a cost-effective approach for commercialized products.

At EMD Millipore, costs to produce hMSCs were analyzed to show the economic benefits of stirred tank bioreactors (e.g. Mobius® CellReady 3L) versus CellStack® plates. This cost analysis only takes into account the upstream process, since the downstream process is the same for both culture vessels. It is calculated that it would cost $69 to produce 10 million cells in a 10 layer-CellStack® plate, while in the Mobius® CellReady 3L it costs $22, which is one-third of the cost of generating the cells in a CellStack® plate (Table 2.3).

Furthermore, the overall cost was broken down to subtypes, such as vessel, media, labor and others, to have a better understanding of the cost (Table 2.4). It is found that the media cost dominates the overall cost, both in CellStack® plates (70%) and Mobius® CellReady 3L (44%), and the savings from Mobius® CellReady 3L primarily can be attributed to the cost of media.

Table 2.3 Cost comparison of 10 million cells in Mobius® CellReady 3L and 10-stack CellStack® Plate.

10^7 cells	Cost	%
CellStack® Plate (10 layers)	$68.78	100.00
Mobius® CellReady 3L	$21.89	31.82

Table 2.4 Cost breakdown analysis.

Item	CellStack® Plates	Mobius® CellReady 3L Bioreactor
Vessel	12%	32%
Media	70%	44%
Labor	15%	9%
Others	3%	15% (include microcarriers)

Because of the pros and challenges of producing large numbers of cells in the multilayer 2D culture vessels discussed above, significant efforts are underway to develop alternatives (such as stirred tank bioreactors, wave bags or fixed-bed bioreactors) to traditional 2D cultures. One promising approach is the expansion of hMSCs on microcarriers in stirred tank bioreactors. This technology will enable the development of a scale-up process in which expansion can be supported in small-scale vessels (e.g. 3 L) and large vessels (e.g. 1,000 L). While these processes are still in development, it can be estimated that a single 1,000 L bioreactor will provide enough cells to produce 1,000 clinical doses, thus microcarrier-based expansion of hMSCs offer an attractive process for allogeneic cell therapy products (Jon Rowley, 2012).

2.4.2 Methodology for Screening of Microcarriers

Most stem cells by nature are adherent cells that need surfaces on which to proliferate, such as ESCs/iPSCs (Jing et al., 2010; Kehoe et al., 2010). This is especially true for hMSCs, since the capability of adherence to a plastic surface is one of their minimal criteria defined by international Society for Cellular Therapy (ISCT) (Dominici et al., 2006). Although it has been shown that hMSCs can be expanded in spheroid suspension culture with serum-free media (Alimperti et al., 2014) while retaining its proliferation and differentiation potential, serial passage studies are still required to confirm its functionality after multiple passages in shake flasks. Dromard et al. (2011) also reported successfully culture of human adipose derived stroma/stem cells as floating spheres in serum-free medium, but again the same serial passage investigation was missing from their study. Because of definition of the hMSC as adherent cells, more publications have been reported to expand hMSCs on various types of microcarriers in bioreactors (Elseberg et al., 2012; Hervy et al., 2014; Jing et al., 2013a, b; Rafiq et al., 2013).

As previously discussed, currently there are multiple types of commercially available microcarriers, varying in backbone, coating, surface charge and area, from different vendors, as shown in Table 2.2. From our experience, each stem cell bank has a very different performance on different microcarriers, especially with respect to their growth kinetics. To identify the optimal microcarriers for each specific cell bank and meet requirements (e.g. animal free requirements for cell therapy), we begin screening microcarriers by investigating hMSCs attachment efficiency, release rate from microcarriers and growth kinetics. To reduce the overall amount of investigation cost and timeline, we:

1) initially screen 11–13 microcarriers from Table 2.3 in static 2D low attachment Petri dishes and choose the 2–4 best performers based on the above-mentioned criteria;
2) further screen these 2–4 microcarriers in spinner flasks and then determine one or two final candidates; and
3) finally verify and/or optimize the final one or two microcarriers in Mobius® CellReady 3L bioreactors.

The whole screening process is schematically summarized in Figure 2.3. However, because most of microcarriers have different surface area per gram, in order to compare cell growth among all microcarriers, an equivalent total surface area and the same amount of hMSCs were seeded in each experiment. From internal experience, we consistently found that a high percentage of viable cells can be recovered from collagen-coated microcarriers or Hillex microcarriers. However, we also found that the percentage of viable cells released from Cytodex® beads is much lower, even though hMSCs growth rate was high on Cytodex® beads.

2.4.3 Process Development in Small-scale Bioreactors (3L)

Unlike the traditional biopharmaceutical process for large biological products (e.g. proteins or vaccine), in cell therapy bioprocess, the cells are the final products so it is critically challenging to obtain maximal biomass of stem cells meanwhile maintaining their purity, identity,

Figure 2.3 Schematic summary of microcarrier screening for a specific stem cell line from initial screening in 2D low attachment Petri dishes to final microcarrier selection in spinner flasks and then proof of principle in Mobius® CellReady 3L Bioreactor (courtesy of EMD Millipore Corporation).

efficacy and safety. It is well known that stem cell products are mixed populations of cells, therefore specifications of purity should be well defined to reflect this. Stem cell potency should also be established by designing of assays to cover both the functional and marker-based studies. The identification of stem cell products should take into account both the specific biomarkers present or absent on the cells and their proliferation capability. Safety involves the development of tests to avoid the contamination of any microbial, endotoxin and mycoplasma, etc. After establishing the guidelines for final product, the key challenge in the bioprocess development is to optimize the operating parameters, such as cell seeding density, microcarrier type and concentration, operation range of pH and dissolved oxygen, and feeding strategy (Kirouac and Zandstra, 2008; Placzek et al., 2009). Given the variability in the setting of these key parameters, it is vital that the optimal operating range for each stem cell product is determined specifically for that product.

From the description within the previous section, each specific microcarrier type can be selected for a stem cell product. The primary screening and selection is performed either in Petri dishes or in spinner flasks; however, these vessels are inadequate to test the bioprocess parameters discussed above. To develop an optimal bioprocess with a reduced cost, the above-mentioned key parameters are first investigated in the small-scale bioreactors, for example Mobius® CellReady 3L, which is a single-use stirred tank bioreactor designed by EMD Millipore for bench-scale cell culture process development.

2.4.3.1 Cell Seeding Density

For adherent cells, the initial seed density of cells per volume of media in the bioreactors is known to influence the cell attachment to the microcarriers and therefore their growth, especially for bone marrow derived hMSCs. The range of seeding density can vary broadly, depending on the specific stem cell bank, the selection of microcarriers and the culture system. A high seeding density (e.g. 10 million cells/L) reduced the lag phase to 1–2 days and reached maximum cell number in a shorter period of time, whereas a low seeding density (e.g. 5 million cells/L) increased the overall expansion fold and total number of cells, but also resulted in a longer lag phase and peaked 3 days later, as shown in Figure 2.4.

The intermittent agitation (2 cycles of 45 min at 25 rpm followed by 45 min without agitation) during the first 3 hours of bioreactor culture

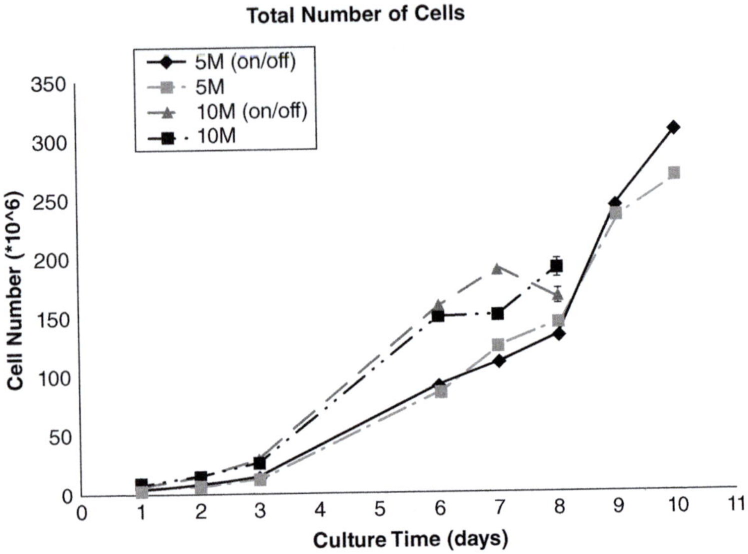

Figure 2.4 Growth of BM-derived hMSCs in Mobius® CellReady 3L bioreactors at different seeding densities while keeping the impellor ON/OFF every 45 minutes for 2 cycles (courtesy of EMD Millipore Corporation).

or continuous agitation (25 rpm) were also investigated. As shown in Figure 2.4, there were no significant differences in overall cell growth between attachment with continuous agitation or intermittent agitation with either 5 million cells/L or 10 million cells/L seeded. The cell attachment efficiency was calculated as the percentage of the attached cells to the seeded cells. It is shown in Figure 2.5 that, over the course of 24 hours, a comparable cell attachment efficiency (70–90%) was achieved both using continuous agitation and intermittent agitation. It should also be noted that under both conditions, the viability of the cells exceeded 90%.

2.4.3.2 Microcarrier Concentrations (Cells:Bead Ratio)

The requirement for an optimal microcarrier concentration at a given cell seeding density may be due to the microenvironment of the microcarriers. Although the microcarriers and cells are well mixed in the bulk bioreactors, each microcarrier is a unique domain and has its own microenvironment, since the number of cells per microcarrier can vary widely. A change in the number of cells per microcarrier

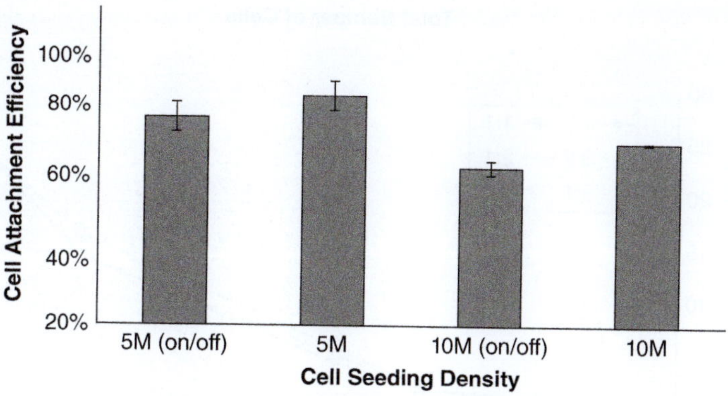

Figure 2.5 Cell attachment efficiency at different seeding densities while keeping the impellor ON/OFF every 45 minutes for 2 cycles (courtesy of EMD Millipore Corporation).

would change the cell distribution on the microcarrier and thus affect the cell growth kinetics.

After the initial cells seeding density was established as described, the impact of microcarrier concentration (or Cells:Bead ratio) was also investigated. Bone marrow-derived hMSCs were inoculated into stirred tank bioreactors with the Cell:Bead ratios of 1:2, 1:1, 3:2, 2:1, 5:1 and 10:1. The cell growth kinetics were studied for a week. As shown in Figure 2.6, there was no difference in total cell number for the first 4 days. Afterwards, the growth rates decreased in 5:1 and 10:1 ratio conditions due to the over-confluency of the cells on microcarriers, which was also confirmed by another group (Goh et al., 2013).

2.4.3.3 Operation Ranges of pH and Dissolved Oxygen

It is well known that the physiochemical environment, such as the gradients of pH and dissolved oxygen (dO_2), affects the expansion of most mammalian cells. However, the specific operating ranges of pH and dO_2 can vary differently for each hMSC bank, depending on their derivation and origins, for example bone marrow or adipose tissue. A broad operating range of pH from 6.8 to 8.0 was investigated for BM-derived hMSCs (data not shown). It was found that cell growth rates were greatly reduced at pH of 6.8 and 8.0. At pH of 6.8, no cell growth was observed in the bioreactors for a week. In the range of 7.0 to 7.8, cell growth curves were all clustered together and did highlight substantial differences.

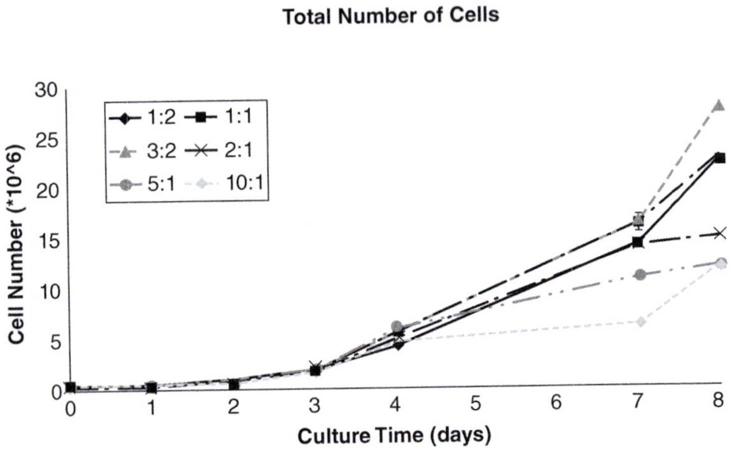

Figure 2.6 Growth kinetics of hMSCs with different cells:bead ratios (courtesy of EMD Millipore Corporation).

Oxygen is important for mesenchymal stem cells, not only because it is one of factors that regulate the metabolism of stem cells, but also it is required to maintain stem cell multipotency. *In vivo*, within the bone marrow, the hypoxic oxygen levels for hMSCs is between 4 and 7% (Grant and Smith, 1963). It has been reported that the hMSCs growth rate can be improved under hypoxic conditions while the multipotency is retained. However, our studies showed no positive effects on cell growth, which was contradictory to that reported previously (Malladi et al., 2006; Wang et al., 2005). The observed difference might be due to the different origin of the cells. Detailed information on pH and dO_2 studies can be found from previous publications (Jing et al., 2013a, b).

2.4.3.4 Feeding Strategy

It is well understood that the concentration of key media components and growth factors, such as glucose, glutamine and b-FGF, and the by-products, such as lactate and ammonia, can affect the growth kinetics of hMSCs (Schop et al., 2009). Glucose and glutamine are required to maintain cellular energy and need to be kept above certain levels. On the other hand, the metabolism products, such as lactate and ammonia, can inhibit hMSCs growth above certain concentrations in the medium. For example, we found that the growth rates of hMSCs were reduced if the concentration of lactate in the media was above 1.68 g/L (Jing et al., 2013a, b) or the ammonia concentration was above 5 mM (data not shown). Then it becomes critical to study when and how the fresh medium should be added to supply new nutrients and dilute the concentration of metabolites. From our experience, the consumption of glucose is an excellent indicator of cell growth; thus we maintain glucose concentration above 0.5 g/L. When the glucose concentration in bioreactors is approaching 0.5 g/L, we will feed either with a normal concentration of glucose media (1 g/L) or high glucose media to bring the bulk concentration of glucose in bioreactors to 1 g/L. We also found that the volume of fresh media that was fed each time is critical. Our previous publication showed that a large volume feed strategy (1 L feed in 3 L bioreactors) could effectively dilute the concentrations of metabolic products below the inhibiting levels and stimulate the growth kinetics of hMSCs (Jing et al., 2013a, b). However, a small volume, multiple feed strategy (0.2 L feed) can dilute the concentrations of the metabolites, but cannot bring them below the critical levels.

2.4.4 Process Development in Medium-scale Bioreactors (50L)

Process developments in medium and large sizes of stirred tank bioreactors are similar to those in 3L scale bioreactors, for example microcarrier concentrations, feeding strategy and operating ranges of pH and dissolved oxygen, provided that the geometric designs between two sizes of bioreactors are similar and comparable, or at least not fundamentally different. Using small-scale bioreactors to develop and optimize the process, which can be applied to large-scale bioreactors, can also well reduce the cost and turnover time.

Using the same process developed from the Mobius® CellReady 3L bioreactors, we were able to successfully scale up the expansion in the Mobius® CellReady 50L bioreactors, as shown in Figure 2.7. In order to compare cell growth kinetics in 3L and 50L, the total number of cells was normalized to total number of surface areas defined as cells per surface area. As shown in the graph, after a 4-day lag phase, hMSCs started to grow exponentially and reached plateau after two weeks. Although cell growth rate was higher in 3L bioreactors during the exponential phase, cells per surface area peaked at similar levels of 30,000 cells/cm^2. Because fresh empty microcarriers were fed at day 7 and day 11, we observed a drop in cells per surface area values at those time points.

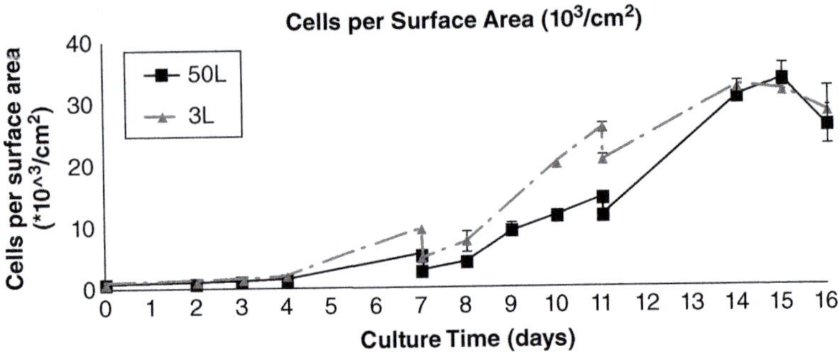

Figure 2.7 Growth kinetics of hMSCs in both Mobius® CellReady 3L and 50L bioreactors (courtesy of EMD Millipore Corporation).

The concentrations for nutrients and by-products were also monitored and compared in 3L and 50L bioreactors with a Bioprofile® FLEX analyzer from Nova Biomedical (Figure 2.8). After 16 days, the glutamine in both bioreactors was almost completely consumed; however, there was glucose (>0.45 g/L) remaining in the bioreactor, indicating that glucose is not the limiting substrate for hMSCs in large volume bioreactors. The concentrations of lactate and ammonia continuously built up in the system; however, neither reached the level of inhibiting hMSC proliferation. More importantly, the consumption of nutrients and build-up of the waste by-products were comparable in both 3L and 50L bioreactors, indicating that Mobius® CellReady 3L bioreactors can be utilized for process development for the larger volume stirred tank bioreactors.

To ensure that cells expanded in 50L bioreactors retain their identity and functionality, hMSCs were harvested to evaluate surface marker expression (Figure 2.9). It is found that cells harvested from two suspension culture systems were both positive for CD90, CD105, CD73 and CD44, but negative for CD14, CD19, CD34, CD45 and HLA-DR, as suggested by Dominici et al. (2006). The differentiation potential of hMSCs harvested from the 50L bioreactors was also evaluated (Figure 2.9). After exposure to adipogenic induction medium, the cells expanded in 50L bioreactors were observed to contain intracellular lipid droplets, which stained positively with oil red O. For osteoblastic differentiation, the cells were shown to stain positive for alizarin red, indicating that some of cells maintained the ability to deposit calcium, thereby indicating an osteogenic capacity. It is also

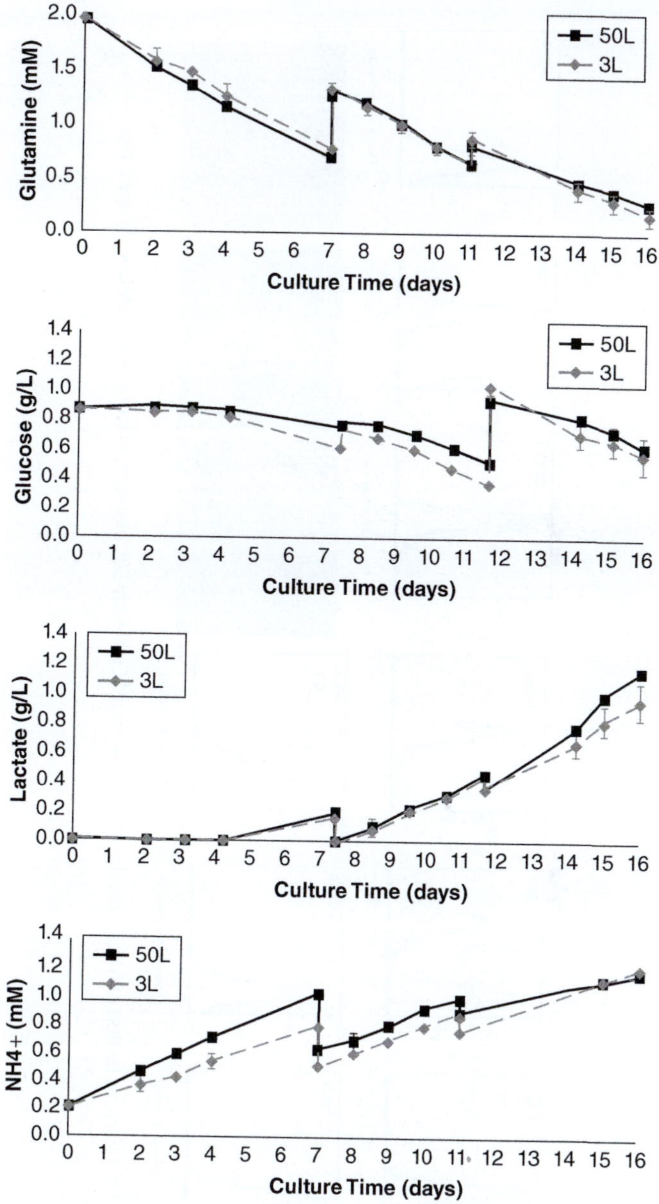

Figure 2.8 The glucose, glutamine, lactic acid and ammonia concentrations in culture over a 16-day culture period (courtesy of EMD Millipore Corporation).

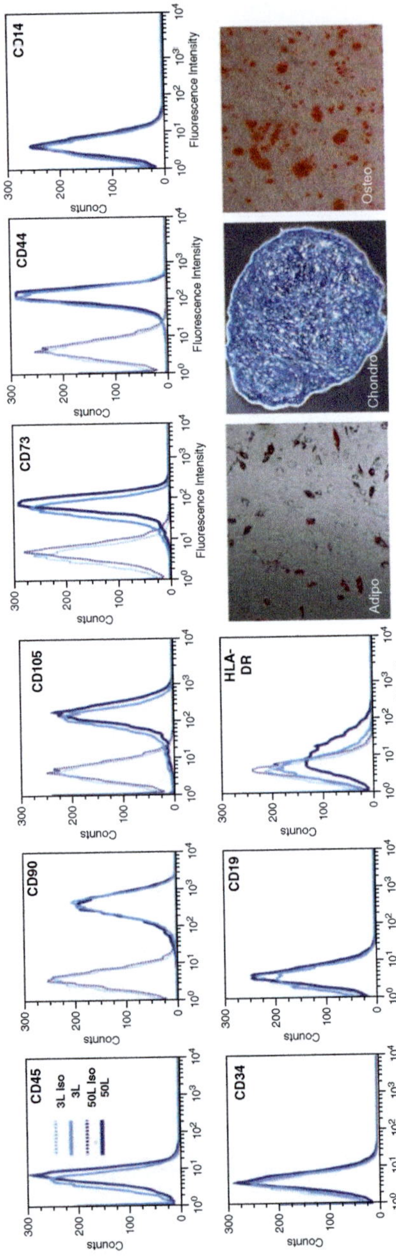

Figure 2.9 Characterization of hMSCs produced from Mobius® CellReady 50L bioreactors (courtesy of EMD Millipore Corporation).

shown that cells retain their capability to differentiate toward chondrogenic lineage.

2.4.5 Case Study for Expansion of Bone Marrow Derived MSCs in Stirred Tank Bioreactors

The existing adherent flat culture process was transitioned to a suspension process. Fresh bone marrow aspirate was used to generate multiple donor banks. Three donors were used in 3L and 50L studies. Based on a survey of the literature, a medium formulation of DMEM, 10% FBS, 2 mM L-glutamine and 8 ng/ml bFGF was chosen for evaluation. Subsequently, we have evaluated serum substitutes and documented equivalent growth.

Two rounds of static and agitated microcarrier screening were performed. Primary readouts were cell attachment and cell recovery, both via cell counts. The top performer in both rounds was collagen-coated microcarriers. Starting with the conditions identified in the spinner flasks, approximately 1,000 c/cm^2 and 15 g/L collagen microcarriers, hMSCs were expanded in 3L bioreactors successfully.

Agitation rate was determined by a cell growth study. Agitation rates of 25, 50 and 100 rpm supported growth at 1 L. Agitation rates of 200 and 300 rpm did not support growth; 50 rpm was chosen as the starting agitation rate for 1 L volume. After 24 hours, more than 60% of the cells were attached to the microcarriers. After expansion, viability was greater than 90%. At days 7 or 8, the agitation rate was increased to maintain the cells-microcarriers in suspension. Nutrients were monitored; glucose and glutamine were depleted, while lactic acid and ammonia accumulated in the reactors. Cell seeding density was investigated at 5 million and 10 million cells per liter of medium using 15 g/L Collagen microcarriers. While 10 million cells could lead to a shorter process, 5 million cells was chosen as the preferred cell seeding density based on fold increase capacity and overall cell yield.

Feed strategy was optimized to increase yield. Three conditions were evaluated:

1) batch process (starting at full volume with all microcarriers);
2) fed batch (starting at low volume and adding medium and microcarriers over multiple feeds) and modified fed batch (starting with 1 L, feeding with 1 L at days 5–7; and

3) feeding with 0.4 L at days 8–10, based on glucose depletion. The modified fed batch performed the best, with a greater than 100% increase in yield.

Cell numbers, nutrients and metabolites are assessed on a daily basis. Cell viability is determined at the end of the run. DAPI staining is performed every 3 to 7 days. Cell surface markers are assessed at the end of the run and occasionally during the run. Tri-lineage differentiation and gene expression are performed occasionally at the end of a run. Karyotyping has been performed for donor 1.

A production run was performed at the 50 L scale and 15 g/L of collagen microcarriers and 50 million cells seeded a 10 L starting volume (directly translated from the 3 L:5 million cells in 1 L volume). Then 10 L, 20 L and 10 L of medium and 15 g/L microcarriers were added over the course of the run. Cell yield was 10 billion cells in a final volume of 50 L. Cell surface markers exhibited the expected results for positive and negative ISCT guidance markers.

In this case study, an existing 2D process was translated to a stirred tank bioreactor expansion protocol. Growth was successful at both the 3 L and 50 L scale.

2.5 Future Directions

The use of stirred tank bioreactors combined with microcarriers provides a robust and cost-effective way to produce a large quantity of hMSCs for allogeneic cell therapy. However, there are several issues that need to be successfully addressed: developing serum-free medium and animal-free raw materials, establishing online assays and downstream harvest processing, and addressing bio-safety concerns.

To date, in most of the clinical trials, cells were expanded in a medium containing various amount of fetal bovine serum (FBS), which causes concerns in clinical application, such as the possibility of transferring animal pathogens to patients and substantial lot-to-lot variability with regional and seasonal supply changes. More importantly, the global amount of cGMP-quality FBS is very limited and may not support successful cell therapy products. Therefore, finding alternatives to FBS is critical for large-scale production of cell therapy products. To address the problem, several approaches have been investigated, such as development of serum-free medium (Al-Saqi et al., 2014, Hervy et al., 2014) or replacement of FBS with human platelet lysate (hPL) (Dudakovic et al., 2014; Shih and Burnouf, 2014). However, the commercial available serum-free medium may not really support every hMSC bank because of the variety of cell origins. For example, it is reported that Mesenchult-XF could provide better growth support for adipose-derived hMSCs than serum-containing medium; however, it does not provide the same performance for bone-marrow derived hMSC for more than 5 passages (Al-Saqi et al., 2014).

So far, among the listed microcarriers in Table 2.2, collagen-coated microcarriers perform the best for hMSC expansion in our hands. However, these microcarriers are coated with porcine gelatin, which poses similar problems as serum-containing medium. Although several reports have been published showing that some commercially available xeno-free microcarriers could support the expansion of hMSCs in bioreactors, the applications are widely reduced with several drawbacks, such as reduced growth kinetics or poor release rate of cells from microcarriers (Kehoe, 2012). More importantly, certain surface treated microcarriers, such as Synthemax® II coated microcarriers, supports hMSCs growth in its designed medium, which limit its application only to certain stem cell banks and the broad selection of medium suppliers. The development of a xeno-free culturing system (including media and microcarriers), online monitoring assays and harvest processes are required before the successful commercial application of cell therapies.

References

al-Saqi, S., Saliem, M., Asikainen, S. *et al.* (2014) Defined serum-free media for *in vitro* expansion of adipose-derived mesenchymal stem cells. *Cytotherapy*, **16**, 915–26.

Alimperti, S., Lei, P., Wen, Y. *et al.* (2014) Serum-free spheroid suspension culture maintains mesenchymal stem cell proliferation and differentiation potential. *Biotechnol. Prog.*, **39** (4), 974–83.

Bourin, P., Peyrafitte, J.A. and Fleury-Cappellesso, S. (2011) A first approach for the production of human adipose tissue-derived stromal cells for therapeutic use. *Methods Mol. Biol.*, **702**, 331–43.

Deskins, D.L., Bastakoty, D., Saraswati, S. *et al.* (2013) Human mesenchymal stromal cells: Identifying assays to predict potency for therapeutic selection. *Stem Cells Transl. Med.*, **2**, 151–8.

Dominici, M., Le Blanc, K., Mueller, I. *et al.* (2006) Minimal criteria for defining multipotent mesenchymal stromal cells. The International Society for Cellular Therapy position statement. *Cytotherapy*, **8**, 315–17.

Dos Santos, F., Campbell, A., Fernandes-Platzgummer, A. *et al.* (2014) A xenogeneic-free bioreactor system for the clinical-scale expansion of human mesenchymal stem/stromal cells. *Biotechnol. Bioeng.*, **111**, 1116–27.

Dromard, C., Bourin, P., Andre, M. *et al.* (2011) Human adipose derived stroma/stem cells grow in serum-free medium as floating spheres. *Exp. Cell Res.*, **317**, 770–80.

Dudakovic, A., Camilleri, E., Riester, S.M. *et al.* (2014) High-resolution molecular validation of self-renewal and spontaneous differentiation in adipose-tissue derived human mesenchymal stem cells cultured in human platelet lysate. *J. Cell. Biochem.*, **115** (10), 1816–28.

Eibes, G., Dos Santos, F., Andrade, P.Z. *et al.* (2010) Maximizing the *ex vivo* expansion of human mesenchymal stem cells using a microcarrier-based stirred culture system. *J. Biotechnol.*, **146**, 194–7.

Elseberg, C.L., Leber, J., Salzig, D. *et al.* (2012) Microcarrier-based expansion process for hMSCs with high vitality and undifferentiated characteristics. *Int. J. Artif. Organs*, **35**, 93–107.

Frauenschuh, S., Reichmann, E., Ibold, Y. *et al.* (2007) A microcarrier-based cultivation system for expansion of primary mesenchymal stem cells. *Biotechnol. Prog.*, **23**, 187–93.

Goh, T.K., Zhang, Z.Y., Chen, A.K. *et al.* (2013) Microcarrier culture for efficient expansion and osteogenic differentiation of human fetal mesenchymal stem cells. *Biores. Open Access*, **2**, 84–97.

Goldstein, S. (1990) Replicative senescence: the human fibroblast comes of age. *Science*, **249**, 1129–33.

Grant, J.L. and Smith, B. (1963) Bone marrow gas tensions, bone marrow blood flow, and erythropoiesis in man. *Ann. Intern. Med.*, **58**, 801–9.

Hervy, M., Weber, J.L., Pecheul, M. *et al.* (2014) Long-term expansion of bone marrow-derived hMSCs on novel synthetic microcarriers in xeno-free, defined conditions. *PLoS One*, **9**, e92120.

Hess, R.D., Weber, F., Watson, K. and Schmitt, S. (2012) Regulatory, biosafety and safety challenges for novel cells as substrates for human vaccines. *Vaccine*, **30**, 2715–27.

Hewitt, C.J., Lee, K., Nienow, A.W. *et al.* (2011) Expansion of human mesenchymal stem cells on microcarriers. *Biotechnol Lett.*, **33**, 2325–35.

Jing, D., Kehoe, D.E. and Tzanakakis, E.S. (2010) Expression of Reg family proteins in embryonic stem cells and its modulation by Wnt/beta-catenin signaling. *Stem Cells Dev.*, **19**, 1307–19.

Jing, D., Punreddy, S., Sunil, N. *et al.* (2013a) Characterization of human mesenchymal stem cells: Expression in a 3-L, single-use, stirred-tank bioreactor. *BioProcess International*, **11**, 30–6.

Jing, D., Sunil, N., Punreddy, S. *et al.* (2013b) Growth kinetics of human mesenchymal stem cells in a 3-l single-use, stirred-tank bioreactor. *BioPharm International*, **26** (4), 28–38.

Jon Rowley, E.A., Campbell, A., Brandwein, H. and Oh, S. (2012) Meeting lot-size challenges of manufacturing adherent cells for therapy. *BioProcess International*, **10**, 16–22.

Jostock, T. and Knopf, H.P. (2012) Mammalian stable expression of biotherapeutics. *Methods Mol. Biol.*, **899**, 227–38.

Kehoe, D., Schnitzler, A., Simler, J. *et al.* (2012) Scale-up of human mesenchymal stem cells on microcarriers in suspension in a single-use bioreactor. *BioPharm International*, **25**, 28–38.

Kehoe, D.E., Jing, D., Lock, L.T. and Tzanakakis, E.S. (2010) Scalable stirred-suspension bioreactor culture of human pluripotent stem cells. *Tissue Eng. Part A*, **16**, 405–21.

Kirouac, D.C. and Zandstra, P.W. (2008) The systematic production of cells for cell therapies. *Cell Stem Cell*, **3**, 369–81.

Loffelholz, C., Husemann, U., Meusel, W. *et al.* (2013) Bioengineering parameters for single-use bioreactors: Overview and Evaluation of suitable methods. *Chemie Ingenieru Technik*, **85**, 40–56.

Malladi, P., Xu, Y., Chiou, M. *et al.* (2006) Effect of reduced oxygen tension on chondrogenesis and osteogenesis in adipose-derived mesenchymal cells. *Am. J. Physiol. Cell Physiol.*, **290**, C1139–46.

Pattappa, G., Heywood, H.K., De Bruijn, J.D. and Lee, D.A. (2011) The metabolism of human mesenchymal stem cells during proliferation and differentiation. *J. Cell Physiol.*, **226**, 2562–70.

Placzek, M.R., Chung, I.M., Macedo, H.M. *et al.* (2009) Stem cell bioprocessing: fundamentals and principles. *J. Roy. Soc. Interface*, **6**, 209–32.

Rafiq, Q.A., Brosnan, K.M., Coopman, K. *et al.* (2013) Culture of human mesenchymal stem cells on microcarriers in a 5 l stirred-tank bioreactor. *Biotechnol. Lett.*, **35**, 1233–45.

Santos, F., Andrade, P.Z., Abecasis, M.M. *et al.* (2011) Toward a clinical-grade expansion of mesenchymal stem cells from human sources: A

microcarrier-based culture system under xeno-free conditions. *Tissue Eng. Part C Methods.*, **17**, 1201–10.
Schop, D., Janssen, F.W., Borgart, E. *et al.* (2008) Expansion of mesenchymal stem cells using a microcarrier-based cultivation system: growth and metabolism. *J. Tissue Eng. Regen Med.*, **2**, 126–35.
Schop, D., Janssen, F.W., Van Rijn, L.D. *et al.* (2009) Growth, metabolism, and growth inhibitors of mesenchymal stem cells. *Tissue Eng. Part A*, **15**, 1877–86.
Schop, D., Van Dijkhuizen-Radersma, R., Borgart, E. *et al.* (2010) Expansion of human mesenchymal stromal cells on microcarriers: growth and metabolism. *J. Tissue Eng. Regen. Med.*, **4**, 131–40.
Shih, D.T. and Burnouf, T. (2015) Preparation, quality criteria, and properties of human blood platelet lysate supplements for *ex vivo* stem cell expansion. *N. Biotechnol.*, **25: 32** (1), 199–211.
Wang, D.W., Fermor, B., Gimble, J.M. *et al.* (2005) Influence of oxygen on the proliferation and metabolism of adipose derived adult stem cells. *J. Cell. Physiol.*, **204**, 184–91.

3

The Effect of Scale-up on Cell Phenotype: Comparability Testing to Optimize Bioreactor Usage and Manufacturing Strategies

Jason Hamilton[1] *and Bart Vaes*[2]

[1] *Athersys Inc, Cleveland, OH, USA*
[2] *ReGenesys BVBA, Heverlee, Belgium*

3.1 Introduction

The application of cellular therapy in regenerative medicine is rapidly expanding, as many preclinical and clinical studies have demonstrated safety and even clinical benefit across a diverse range of applications. The approval of autologous (e.g. ChondroCelect; HoloClar®, Provenge®) and allogeneic therapies (e.g. Prochymal®), and the growing number of ongoing clinical trials are strong indications that additional applications of cellular therapies will become available over the next decade. While for most applications the cells are not yet brought to market, novel opportunities are already explored in clinical trials, in which iPS and engineered stem cells are being tested. Currently, the therapeutic cell products under development are expanded on traditional 2D-based platforms in media that contain bovine serum. The field faces significant manufacturing challenges that need to be overcome before product commercialization and entrance to large-scale markets. In this chapter, we focus on cell characterization related to manufacture of bone marrow-derived adherent stem cells to guarantee product consistency upon scale-up of product manufacturing.

The increase in therapy-driven stem cell research over the last decades, and the identification and isolation of human stem cells that can be expanded *in vitro*, have led to translational application of cells into

Bioprocessing for Cell Based Therapies, First Edition. Edited by Che J. Connon.

the clinic. Bone marrow-derived adherent stem cells have been of particular interest based on their ease of isolation from adult donors and *in vitro* expansion capacity using standard tissue culture methods, reagents and culture vessels. Preclinical studies have suggested that mesenchymal stromal cells (MSC) may provide benefit within a wide variety of diseases and injuries, potentially due to the multimodulatory effects that the cells exert on surrounding cells and/or tissues. For instance, bone marrow-derived cells have been shown to migrate to injury sites, to exert pro-angiogenic effects, and to modulate the immune system via suppression of activated T-cell proliferation. There are currently more than 350 clinical studies documented (Bersenev, 2015). The specific bone marrow-derived cells being most used are MSC, and are currently under investigation for safety and efficacy in a broad range of therapeutic areas including cardiovascular, neurologic, immunologic and tissue transplantation applications. Currently, 40 clinical studies using MSC as intervention are in Phase III/IV and 6 of those studies have been completed (ClinicalTrials.gov, targeted search "Intervention: mesenchymal"). Given the increase in number and relative success in trials, growth of the cell therapy market treating large numbers of people can be expected over the next decade. Consequently, the rapidly growing cell therapy industry needs significant process improvement and cell manufacturing scaling in order to meet the large cell numbers that will be required for market entrance and commercial success. Thorough cell characterization is needed to exclude significant variation in product identity, activity or consistency, after such changes in manufacturing.

3.1.1 Cell Characterization in the Development Path

Cell characterization assays used in comparability testing should be developed around three central aspects of product identity:

1) consistency in identity, as demonstrated by specific measurements observed in past manufacturing runs or process changes;
2) maintenance of unique product identity/distinctions from related products or similar, but different cell types; and
3) demonstration of consistent and therapeutically relevant product potency.

The first aspect of product identity depends on prior demonstration of some induction of variability in a specific measure of product identity. Such observation of variability may occur in response to specific process changes, or may serve as a marker of production specifications that are out of range (i.e. excessive passage number). Monitoring of such indicators of variability should help to maintain manufacturing specifications that result in optimal product. Maintenance of product distinction from similar products or related cell types serves a similar purpose, but may rely on known markers of cell type specificity (as opposed to requiring historical measures of variability). Monitoring of specific cell type markers will ensure lack of product "drift" towards related identities, potentially serving both potency and intellectual property concerns. Finally, and most importantly, characterization of cell product must rely on some biologic understanding of underlying potency within the intended therapeutic area.

Product comparability ultimately requires maintenance of equivalent potency, and understanding product attributes that contribute to said potency will allow for successful assessment of comparability. Product attributes that affect potency may be rather direct – secretion of specific levels of protein(s) that drive a biologic effect for example – and may be directly measurable *in vitro*. Other product attributes affecting potency may be more indirect. Following the prior example, secretion of specific protein(s) responsible for the potency of the product may require other factors for said potency. Cell size and cell surface receptor expression may have significant influence on homing/biodistribution, cell–cell interactions, etc. that may significantly impact the ultimate location(s) at which the specific protein(s) are secreted by the cell product. Cell division rate may impact the ultimate number of cells that are present to secrete the protein(s) responsible for potency. Specific differentiation and appropriate long-term engraftment may be required for product potency. Many other product attributes could similarly impact ultimate product potency, and thus may be equally important to monitor in the assessment of comparability.

Successful integration of appropriate product characterization assays during clinical and commercial product development is a complex process requiring harmonization of discordant assay considerations. From the scientific perspective, the more sensitive and informative the assay, the more telling the resulting data will be regarding the subtleties of product identity. Conversely, from the commercial drug manufacturing perspective, the more simple and direct the assay, the more feasible it will be to implement as a validated process, in terms of assay practicality (time/effort required to complete), transferability (across multiple assay performers, multiple sites, outside contract organizations, etc.), and robustness (ease of interpretation of results, ease of assay validation). Ultimately, inclusion of assays representing both ends of this spectrum will be required for successful assessment of product comparability/consistency. In addition, these different types of assays will serve slightly different purposes in successfully demonstrating product comparability/consistency.

Provision of a thorough and informed assessment of cell therapy product consistency is of primary concern to regulatory agencies. This regulatory concern requires the cell therapy manufacturer to provide datasets upholding that:

- product lots are consistent; and
- the data presented to illustrate consistency, and the assays performed to collect said data, are truly relevant and sufficient to illustrate product consistency.

The adoption of the above described spectrum of assays can be used to meet these requirements. Performance of more complex "scientific investigative" assessments (microarray analysis of gene expression, secretome analysis of protein secretion, miRNA profiling, mass cytometric [CYTOF] survey of surface protein expression, etc.) can generate large, informative datasets allowing for more thorough and detailed understanding of product phenotype. These profiling datasets can similarly inform appropriate development of simpler, more robust assays, measuring a single factor or component of the product, that

Table 3.1 Cell comparability in the development path.

Analytical tools support consistency in product manufacturing and process change towards commercialization
Donor consistency important for implementing master cell bank strategy
Large-scale manufacturing moves towards extended population doublings and needs to ensure consistent safety and potency
Ensuring distinction between competing products in adherent stem cell space – enforcing and growing IP boundaries
Building a regulatory strategy for implementing major process changes (xeno-free media; alternate bioreactor work; driving down COGS)

may serve as a valuable, routine assessment of some central aspect of product identity (a "Critical Quality Attribute", or CQA). Ultimately, development of a defined set of validated, robust, routine release assays that are believed to encompass a defined set of CQAs – all scientifically substantiated by supportive data obtained through non-validated "R&D" assessments – will help to assure regulators that product consistency is thoroughly understood and appropriately monitored.

3.1.2 The MultiStem® Allogeneic Cell Therapy Product: Mechanisms of Benefit and Target Cells Numbers

To illustrate the required cell numbers for clinical studies and mechanisms of action, the MultiStem® allogeneic cell product will be taken as an example. The MultiStem product, as developed by Athersys Inc., is a clinical product based on expansion protocols for Multipotent Adult Progenitor Cells (MAPC), which are non-hematopoietic adherent cells derived from the bone marrow of healthy adults (Boozer et al., 2009; Jiang et al. 2002). Expanded MAPC cells have a unique phenotype that discriminates them from mesenchymal stem cells, with significantly enhanced proliferation capacity as one important advantage for MAPC. Human and rodent MAPC are non-immunogenic, lack MHC II expression, and therefore do not induce a proliferative response when co-cultured with allogeneic T-cells and significantly reduce T-cell proliferation when responder T-cells are stimulated with allogeneic irradiated stimulator cells (Jacobs et al., 2013; Kovacsovics-Bankowski et al., 2009). The product can be kept in frozen form, thus making the MultiStem product an adult-derived "off-the-shelf" stem cell product platform, with therapeutic product available at the time of need and usable for any potential recipient without any requirement for tissue matching.

The therapeutic benefit of MAPC/MultiStem cells has been shown in diverse pre-clinical animal disease models. For instance, the cells improve tissue regeneration and function in cardiovascular and neurological disorders, including acute myocardial infarct, ischemic stroke, traumatic brain injury, spinal cord injury and ischemic limb injury (Aranguren et al., 2008; 2011; Busch et al., 2011; Mays et al., 2010; Ryu et al., 2011; van't Hof et al., 2007; Walker et al., 2010, 2012). Also, MAPC have been employed to prevent graft versus host disease (Kovacsovics-Bankowski et al., 2009) and mediate allograft survival after solid organ transplantation (Eggenhofer et al., 2013).

Based on these varied findings, MultiStem cells have been tested or are currently being assessed in multiple clinical studies in cardiovascular, neurological, inflammatory and immune disease areas. Safety studies (Phase 1 clinical trials) using MultiStem as an adjunct in allogeneic bone marrow transplantation for prevention of GvHD (Maziarz et al., 2012) and acute myocardial infarction (AMI) have been completed (Penn et al., 2012). Currently, MultiStem cells are being tested in:

- a Phase 1 safety study for immunomodulation therapy after liver transplant;
- a Phase 2 randomized multicenter sham-controlled study in patients after heart attack; and
- a Phase 2 randomized, double-blind, placebo-controlled, multicenter study of ischemic stroke patients with treatment groups receiving doses of 400 million or 1,200 million cells per patient (Hess et al., 2014).

The therapeutic benefits of MAPC are multimodal, and identified underlying mechanisms include the pro-angiogenic and anti-apoptotic effects mediated via trophic support (Aranguren et al., 2007, 2011; Lehman et al., 2012), as well as their potent ability to modulate the immune system via suppression of T-cell proliferation or shifting the balance in macrophage M1-M2 polarization (Kovacsovics-Bankowski et al., 2009; Reading et al. 2013; Walker et al. 2010). In preclinical models of ischemic stroke and traumatic brain injury, cell labeling revealed that after intravenous injection, the MAPC distribute to the spleen, subsequently preventing loss of splenic mass and splenocyte apoptosis, and eventually result in reduced blood brain barrier permeability. The importance of the spleen in the regulation of this process, and the demonstration that MAPC accumulate in the spleen after injection in these models of CNS injury, illustrates that cell homing may also be a CQA necessary for therapeutic benefit.

The use of autologous as well as allogeneic cell products requires *in vitro* expansion to achieve sufficiently high cell numbers; and, dependent on the clinical need, different strategies may be followed. When cells are administered within a short period after disease occurrence, as is the case for the use of MultiStem being tested in ischemic stroke, allogeneic cells have an advantage over autologous cells. However, before safe application as an allogeneic cell product to patients, the cells need to be isolated, expanded, quality tested and stored in order to provide sufficient cells with therapeutic activity.

For the MultiStem product, a manufacturing process has been developed that utilizes a 2-step strategy, in which working cell bank campaigns for clinical product are obtained from a master cell bank. This strategy enables production of cells for more than 100,000 doses from a single donor. The creation of master cell banks and production campaigns from these banks are done by a contract manufacturing organization on 10-layer cell factories. The dose of one production unit contains 180 million cells, and per production run of about 40–50 clinical doses can be generated.

Although this strategy allowed for consistent production of cells for current clinical studies, larger production platforms will be required to meet the product demands for clinical market entry (Table 3.2). It is therefore important to consider alternate production strategies and move towards technologies with significantly increased cell expansion surface area (Rowley et al., 2012).

Different types of bioreactors are being explored for future MultiStem production, including the Quantum Cell Expansion System (TerumoBCT), which is a closed automated culture system containing a disposable hollow-fiber bioreactor of 2.1 m^2 surface area. The reactor is connected to a sterile closed-loop computer-controlled media perfusion platform and gas exchangers (Antwiler et al., 2009), on which 1 billion cells can be harvested, representing 5 doses of current product formulation. Nevertheless, such 2D technologies are not sufficient to achieve trillions of cells in a single run.

Table 3.2 Example of required cell numbers for several cell therapy applications.

	GvHD prophylaxis	AMI	Stroke
Patients per annum	10,000	200,000	500,000
Treatment: cells per patient (million cells)	400–1,600	50	500–1,000
Cell requirements per annum (trillion cells)	4–16	10	250–500
Run size	Minimum feasibility: ~50 billion cells run		Larger runs required

3.2 Challenges in Cell Product Development

3.2.1 Effect of Large-scale Expansion on Stem Cell Properties

In order to proceed with large-scale Phase III studies, and enter the market after product approval, the field has to overcome a number of challenges to be able to manufacture the cell numbers required with sufficient consistency. This is only possible after scale-out (e.g. for autologous therapies), or up-scaling or moving to novel expansion platforms (planar, packed bed, hollow fiber, microcarriers) for allogeneic products. Such bioreactors can be scaled up to volumes of 1,000 liters, and gas exchange, nutrient supplementation and pH can be precisely controlled. Suspension cultures are likely the only possibility to achieve the trillions of cells required for allogeneic product manufacturing. In addition, downstream processing in functionally closed systems is key to recovering the therapeutic cells at high cell concentration and purity levels (Pattasseril et al., 2013). Volume reduction, cell washing, formulation, filling into vials, freezing and storage are all aspects of downstream processing that may affect product purity and quality.

For adherent cells, the area-to-volume ratio can be drastically increased by using microcarriers in stirred bioreactors. A diversity of microcarriers is already commercially available and can be classified into different categories, namely:

1) positively charged non-porous smooth;
2) collagen coated
3) ECM coated;
4) non-charged;
5) macroporous; and
6) silica-weighted (Chen et al., 2013).

Given that these microcarriers are composed of different types of matrices (e.g. dextran, cellulose, polystyrene, glass, gelatin) and may contain various types of surface coatings (e.g. collagen, fibronectin, Synthemax®), it is not unexpected that the type of microcarrier strongly influences cell adherence and stem cell qualities.

The microenvironments of the 3D structures are unlike planar cultures and cells experience not only physical forces such as shear stress in stirred reactors, but also cytoskeletal tensions because of the curved microcarrier surface. When cultured on microcarriers, cells tend to form aggregates, and such changes in the microenvironment of cells impacts matrix deposition and the cellular skeleton (Zeiger et al., 2012).

Moreover, as stem cell commitment is influenced by surface tension, mechanical cues and cytoskeletal tension can influence commitment between adipocyte and osteoblast (McBeath et al., 2004). In accordance, MSC cultured on collagen-coated microcarriers may undergo spontaneous osteogenesis and lose adipogenic and chondrogenic differentiation capacity as a result of increased cytoskeletal tension (Tseng et al., 2012). On the other hand, soft tissues can favor adipogenic differentiation (Galateanu et al., 2012), and stemness and differentiation can be modified through manipulation of conditions that promote low cytoskeletal tension (Eyckmans et al., 2012; Zhang and Kilian, 2013). The significant potential for microcarrier expansion to affect differentiation potential was demonstrated by comparing human fetal MSC expansion on 2D and microcarriers (Goh et al., 2013). Although growth kinetics and cell surface markers were similar, the microcarrier expanded cells showed enhanced *in vitro* osteogenic capacity and led to more ectopic bone formation after injection in mice.

Although losing stemness and formation of osteoblasts on microcarriers is disadvantageous for stem cell production, the osteogenic induction can also be used as bone grafts for tissue engineering. Six weeks of dynamic culture of adipose derived mesenchymal stem cells on microcarriers induced osteogenic phenotype stronger than in 2D cultures (Declercq et al., 2013). In this way, advantage can be taken of such alterations; but for scaling of stem cell manufacturing, it is necessary to maintain the stem cell phenotype which in most cases will be proliferation of the undifferentiated cell.

It has become obvious that, not only are size and shape of the microcarrier important, but microcarrier surface chemistry is also a determinant of cell phenotype and yield. For instance, human embryonic stem cell (hESC) yield is improved by a Matrigel coating, but in a cell line dependent manner (Chen et al., 2011). Unlike other cells, these cells may require a laminin or Matrigel coating of microcarriers to replace the feeder cells that are traditionally required for embryonic stem cell expansion. A comparison between two different hESC lines revealed that both lines showed different expansion properties and maintenance of pluripotency when expanded on laminin versus Matrigel coated microcarriers. For one of the two lines, growth and pluripotency was maintained on Matrigel, but not on laminin, likely due to the gelatinous nature of Matrigel that may have protected cells from mechanical stress.

There are already numerous examples in literature of MSC cell expansions on different microcarriers, but for each case different types of microcarriers with different coatings may need to be tested, since the exact properties of how cells perform on a microcarrier may even be cell line dependent. Hence, the choice for a particular microcarrier depends on multiple factors including cell type, attachment, cell growth, pluripotency and harvesting efficiency, which will eventually determine therapeutic applicability.

3.2.2 Serum-free and Xeno-free Media Development

The scale of manufacturing is not the only factor that is changing for future therapeutic application. Also the medium may change when large cell numbers are needed at market entrance, in particular for disorders where high cell doses are required. Current expansions are often done in media containing bovine serum, which has a number of drawbacks. Apart from ethical, safety and consistency issues, the availability of serum will become limited if it is required for the culture of cells up to the scales mentioned above (Brindley et al., 2012). Serum lot-to-lot variation may also have significant impact on product consistency. These arguments strongly favor the use of alternative media in which bovine serum is either replaced by human serum or plasma lysate, or (chemically defined) xenobiotic-free media, of which several that are designed for MSC expansion are commercially available.

Functional alternatives for animal serum replacement in MSC culture are human platelet lysate (HPL) and autologous (donor) serum (AS) and several groups have reported the successful expansion of MSC in HPL or AS, thereby maintaining cell characteristics (references). Adipose-derived stem cells expanded in pooled human platelet lysate showed enhanced growth speed compared to cells in FBS, maintained ISCT flow cytometry and trilineage differentiation criteria, and had similar pro-angiogenic capacity as FBS expanded cells (Trojahn Kolle et al., 2013).

The first published studies are appearing in which xeno-free microcarrier stirred suspension systems have been used to expand different therapeutic cell types, including human induced pluripotent stem cells (Badenes et al., 2015; Fan et al., 2014) and MSC (Carmelo et al., 2015; Dos Santos et al., 2014; Santos et al., 2011). In the study by Dos Santos et al., ISCT criteria were maintained, but potency assays were lacking.

Another component of cell therapy manufacturing in which cell characterization is vital is in developing storage methods for the stem cell product. There is ongoing debate regarding the relative activity of cryopreserved versus fresh stem cell products. For instance, cryopreserved cells post-thaw differ from MSC freshly taken from culture when administrated *in vivo* in mice, as thawed cells were not detected in the lung in contrast with fresh MSC (Chinnadurai et al., 2014). However, metabolic activity was similar between the cells, but disrupted actin polymerization during the freeze-thaw process was suggested to be the underlying cause. It has further been suggested that the inhibitory effect on T-cell proliferation by cryopreserved MSC is impaired when compared to MSC out of culture (Francois et al., 2012). A similar result was found in a study by Moll et al. (2014) and these authors argued that therapeutic effects may be affected by freeze-thawing, as a trend was seen that very low passage MSC led to better response than thawed cryopreserved MSC in a GvHD patient. It must be noted, however, that in this latter study low passage cells were compared to higher passage cells. Also, the fact that these findings have not been widely replicated, could result from suboptimal cryopreservation and/or thawing methods, and such differences have for instance not been observed with MultiStem cells (data not shown).

Although fresh cell administration may be applicable for autologous applications, it is not a feasible approach for allogeneic off-the-shelf products. Nonetheless, the suggestion of altered activity with cryopreserved and thawed cells is of potential importance, and illustrates that properly chosen potency assays will be of great significance not only after cell expansion, but also in the development of the cryoformulation and the freeze-thaw procedure.

Despite many reports on successful stem cell expansion on microcarriers, serum-free medium or a combination of both (Dos Santos et al., 2011), care must be taken in each case that the stem cell phenotypes remain unaltered and that therapeutic activity is not lost.

3.3 Stem Cell Characterization

3.3.1 ISCT Requirements

Different characterization parameters can be applied to characterize a stem cell product, including physicochemical characterization, biologic characterization, potency and comparability testing, and stability testing (Bravery et al., 2013). Physicochemical characterization refers to cell morphology, cell numbers and phenotypic markers such as gene and protein expression. Biologic characterization relates to biologic function such as cell growth, differentiation, proliferation, etc., while potency assays are important to show the biologic activity of the cell product. For a consistent description of MSC, the International Society for Cellular Therapy (ISCT) has proposed minimal criteria to define these cells (Dominici et al., 2006). MSC are defined as cells that:

- adhere to tissue culture plastic;
- have a specific set of surface marker expression; and
- show multipotent differentiation.

For surface markers, more than 95% of the MSC population must demonstrate expression of CD73, CD90 and CD105, while CD45, CD34, CD14 or CD11b, CD79α (or CD19), HLAII must be less than 2%.

Multipotent differentiation is demonstrated when cells can undergo *in vitro* differentiation towards the osteogenic, adipogenic and chondrogenic lineages. Although these criteria are of great importance to define the MSC phenotype after isolation from different tissues, or to exclude remaining fractions of, for instance, hematopoietic cells, they may be insufficient to demonstrate cell product consistency after changes in the development path. For instance, the absence of hematopoietic cell surface markers does not rule out differentiation of MSC during the expansion procedure. Also, the clinical relevance of differentiation into osteoblasts, adipocytes and chondrocytes is not immediately clear, while important regenerative characteristics such as cell migration, pro-angiogenic activity and immunosuppression are not taken into account by these criteria. Finally, this panel does not necessarily discriminate between competing products. For instance, both MAPC and MSC are derived as adherent populations from bone marrow aspirates, which consist of a complex pool of stromal as well as hematopoietic cells. MAPC would likely meet most or all of the ISCT criteria, given the criteria's non-specific applicability to a variety of stromal cell populations. And, both MAPC and MSC can have comparable potent immunosuppressive effects as measured by *in vitro* T-cell proliferation (Reading et al., 2013). However, the expanded cells exhibit unique RNA expression profiles, resulting in unique pro-angiogenic and neuroprotective effects by MAPC in animal models (Boozer et al., 2009; Mora-Lee et al., 2012; Roobrouck et al., 2011).

3.3.2 Potency Assays

Detailed biological characterization is required to know your cell product and to guarantee product consistency across multiple donor isolations or different manufacturing runs.

As indicated above, pro-angiogenic activity, immunomodulatory capacity and migration are important stem cell properties for therapeutic benefit. A more elaborate assay, or panel of assays, is therefore essential to guarantee product consistency and activity. Such potency assays are relevant to assess the therapeutic efficacy of stem cell products (Bravery et al., 2013). The FDA definition of potency testing for cellular and gene therapy products is:

> ...the specific ability or capacity of the product, as indicated by appropriate laboratory tests or by adequately controlled clinical data obtained through the administration of the product in the manner intended, to effect a given result (Guidance for industry, Potency tests for cellular and gene therapy products).

The advantage of potency assays is that the therapeutic mode of action is the basis for the assay. At the same time, however, this "advantage" may introduce a complexity to the assays since these mechanisms may be of a complex nature. For instance, pro-angiogenic and immunomodulatory capacity depend on the presence of endothelial or T-cells respectively in the assay system. In the tube formation assay, the pro-angiogenic effect is quantified as the capability of secreted factors in cell conditioned medium to induce the formation of capillary structures of endothelial cells. For T-cell suppression, a so-called mixed lymphocyte reaction is often used, in which T-cells (in a population of PBMCs) are induced to proliferate, while a possible suppressive effect of therapeutic cells is tested in a dose-dependent manner. Though such qualified assays work well during the research and development phase, they may be too complicated as validated assays for clinical lot release. For instance, as additional cell types are required, a well-chosen, consistent source needs to be defined for testing. HUVECs can be used as an expandable endothelial cell line in an angiogenic potency assay, but for an immunosuppressive assay,

PBMCs are required of which responsiveness may vary strongly from donor to donor. Also, tube formation needs to be analyzed in a 3D matrix, for which often Matrigel is used; and as a complex gelatinous matrix of heterogeneous secreted proteins, it may introduce significant variability or hamper the robustness of the assay.

Potency assays may be simplified by replacing them with so-called "surrogate potency assays", in which the therapeutic activity is determined in an indirect fashion. An illustrative example is the measurement of cytokine expression to predict angiogenic activity (Lehman et al., 2012). Immunodepletion and adding back of the cytokines Vascular Endothelial Growth Factor (VEGF), chemokine ligand 5 (CXCL5) and Interleukin 8 (IL-8), revealed a correlation between these protein levels in MultiStem conditioned media and angiogenic activity. The ELISA-based measurement can therefore be argued as a surrogate measure of potency, without the need for an additional cellular component within the assay or materials of inconsistent composition. For Prochymal, an MSC product developed by Osiris, expression of TNFR1 is used as a surrogate potency assay to establish potency for immunosuppressive properties (Pritchett and Little, 2012).

Ideally, a (surrogate) potency assay is not only based on the MOA, but also correlates with *in vivo* activity, which has been accomplished for different cell therapy products such as ChondroCelect, an autologous chondrocyte product for cartilage repair, and for AMR-001, an autologous bone-marrow derived cell product for cardiac repair (Bravery et al., 2013). By correlating gene expression profiles of *ex vivo* expanded chondrocytes with their ectopic cartilage formation capacity after intramuscular injection in mice, markers were identified which formed the basis of a qPCR based potency assay for ChondroCelect. The potency assay for AMR-001 is based on the ability of the cells to migrate to the damaged region where SDF-1 is produced. The *in vitro* mobility of CD34+ cells in a gradient of SDF-1 correlated with clinical benefit in a Phase 1 clinical study, thus for this product an *in vitro* migration assay serves as a potency assay.

3.3.3 Omics Screens for Therapeutic Stem Cell Characterization

The example of ChondroCelect mentioned above underscores the importance of "omics" analyses in cell characterization, since the application of gene expression profiling led to the identification of markers that formed the basis of a product potency assay. However, profiling techniques are also strong tools for cell identity testing during the development phase, in order to compare global gene expression profiles after adaptation of expansion methods. That expression profiling data can be combined to extract genes for identity testing as well as potency testing is illustrated by a study in which clinical grade mature dendritic cells (DC) were analyzed from different donors and after different manufacturing techniques (Castiello et al., 2013). By comparing mature DC versus immature DC and monocytes, a subset of genes was identified specific for the clinical product. Variability analysis on this subset revealed 29 genes that showed lowest manufacturing, intra-individual and inter-individual variances, and can thus be considered as markers for identity testing. On the other hand, genes showing high variability were detected, including cytokines that may serve as candidates in surrogate potency testing.

Expression profiling has been used to demonstrate similarity after different culture methods such as serum-free medium (Chase et al., 2012) or different isolation methods from bone marrow (Otsuru et al., 2015). It is even more interesting, however, to apply such data in an iterative manner to optimize the expansion conditions. By looking at functionally related over-expressed genes in microarray data, so-called Gene Ontology Analysis, it was found that reduced proliferation of blood-derived colony forming cells in serum-free medium was correlated with marked up-regulation of lipid biosynthesis (Zeisberger et al., 2010). Such data analysis can provide a path to precise optimization of expansion media with lipids, but in other cases may reveal insight into the use of growth factors or other media constituents.

In addition to mRNA profiling, other profiling techniques may be applied for identity testing of therapeutic cells. Epigenetic stability can be a readout for cell maintenance during expansion, as different populations of stem cells are characterized by different epigenetic signatures and the number of differentially methylated genes of closely related stem cells may be small (Aranda et al., 2009). DNA methylation screening is a method to demonstrate epigenetic stability of the cell product during process development changes. Differential methylation is seen when comparing cells from young or old donors, and changes in methylation of CpG sites may occur during *in vitro* passaging (Bork et al., 2010). For instance, the promoter CpG methylation state of a selected panel of cancer genes was preserved during expansion of MSC in Fetal Bovine Serum as well as autologous serum (Dahl et al., 2008). Also serum-free passaging of MSC led to changes in global DNA methylation and histone acetylation due to altered DNA methylase activity and Histone deacetylase activity respectively (Zhu et al. 2015a, b). However, as the correlation between differential gene methylation and gene expression is not always clear, it is difficult to understand whether altered epigenetic profiles affect cell function and therapeutic efficacy. Different studies report that reducing DNA methylation levels in MSC by expanding the cells in the presence of small molecules can influence the immunomodulatory activity and possible therapeutic effects against immune-related disorders (Lee et al., 2015; Teklemariam et al., 2014). Such results underscore that drastic changes in epigenetic profiles alter cellular functions. Given the non-specific effect of global inhibitors of DNA methylation, it remains to be demonstrated whether such manipulations are useful to enhance therapeutic potency, and extensive cell characterization and animal studies are required to demonstrate efficacy and safety.

Taken together, stem cell therapy holds great promise, but the industry needs to apply significant changes in their manufacturing procedures to meet market requirements. This is only possible when extensive cell characterization is done during process development, optimization and clinical production (Table 3.3).

3.4 Next-generation Stem Cell Development

It should be obvious that potency assays are of great importance to assess the possible therapeutic capacity of next-generation stem cell products. Regenerative medicine enters a new era with the recent first clinical trial in which induced pluripotent cells (iPSC) were created from the skin of a patient, after which the iPSC were induced to differentiate into a sheet of retinal pigment epithelium and subsequently the retinal tissue was implanted into the eye as a remedy for macular degradation (Cyranoski, 2014). In addition, clinical testing of genetically-modified cell products is becoming increasingly common. As development moves forward, novel cell characteristics may be seen. It has been reported that serum-free expanded cells secrete increased levels of angiogenic factors leading to enhanced angiogenesis *in vivo* (Oskowitz et al., 2011). But cells may also be genetically engineered to demonstrate such enhanced potency, which may not

Table 3.3 Overview of possible cell characterization assays.

Assay Category	Assay Readout	Methodology	Considerations
Cell identity	Surface markers	Flow Cytometry	Panel required by ISCT. Sufficient, or are more markers needed? Need proper controls (cells and technology)
	mRNA, miRNA	qPCR	How many genes/targets to be tested? What is the biological rationale of a particular RNA? Only in development path to add scientific rationale to validated release assays. Many different platforms
		Array technology	
	Epigenetic profile	Next-generation sequencing	Data analysis is complex. Not a clear correlation between epigenetic profile and cell phenotype
Cell biology	Viability	Life/dead markers	Important assays for thaw/freeze process
	Adhesion		ISCT criteria.
	Differentiation	Osteogenic Adipogenic Chondrogenic	All relevant for therapeutic activity?
	Cytokine expression	ELISA, alphaLISA	Frequently proposed as surrogate potency assay. Correlation with *in vivo* biology needs to be determined.
Cell potency	Immuno-suppression	Mixed lymphocyte reaction	Relevant for therapeutic activity. Impacted by PBMC donor variation
	Angiogenic activity	Tube formation assay	Dependent on matrigel, risk of lot-to-lot variation
	Migration	Scratch assay	Easy method
		Transwell filters	Migration can be measured towards chemokines

only be related to function, but also to migration and homing (Keung et al., 2013).

Not unexpectedly, over-expression of growth factors can lead to significant changes in cellular phenotype in a growth-factor dependent manner. The strength of a good assay panel to address the value of such effects is demonstrated by a study in which expression of different growth factors were enhanced in MSC (Fierro et al., 2011). While PDGF and TGFbeta significantly altered MSC proliferation and differentiation, this was not seen after over-expression of VEGF. Interestingly, VEGF over-expression led to enhanced blood flow restoration after injection in a rodent hind limb ischemia model, providing proof of enhanced therapeutic activity while maintaining standard cell characteristics.

Genetically-modified cell products, including genetically-modified MSCs, have progressed to clinical evaluation. SanBio Inc. is running a Phase I stroke trial with SB623, which are Notch intracellular domain-over-expressing MSC that are reported to exhibit enhanced expression of paracrine factors such as IL-6, IL-8 and MCP-1. The cells are described to have neurosuppressive properties and are in clinical studies for stroke and in preclinical testing for Parkinson's disease and spinal cord injury. *In vitro* comparability testing showed that the cells maintain equivalent immunosuppressive capacity in a mixed lymphocyte reaction compared with typical MSC (Dao et al., 2011), but that conditioned medium of SB623 had higher proangiogenic activity than CM from non-engineered MSC, although the effect was dependent on type of angiogenesis assay and it is not clear if this extends to *in vivo* effects (Dao et al. 2013).

Cellular therapy development continues to expand rapidly, and it should be clear that a sound assay panel that includes data-supported identity and potency tests is obligate for successful development of novel stem cell therapies. Despite the demonstration of biological activity of genetically-modified cells, it remains unclear what the consequences are for larger-scale xeno-free culture, since serially passaged hMSC progressively lose replication potential and multipotency. In general, it is unclear what the correlation is between differentiation potential, marker expression, *in vitro* immunosuppression, *in vitro* pro-angiogenic activity and therapeutic activity. Also, differences in donor cell material are not fully comprehended. In a comparative analysis between donor age, gender, marker expression, *in vitro* differentiation and *in vivo* bone formation capacity, it was shown that marker expression correlated with *in vivo* bone forming capacity, but no clear correlation was seen for donor age (Siddappa et al., 2007). A more extensive study revealed no correlation between donor age and differentiation potential *in vitro*; trophic factors were not determined by age, gender or marker expression; CD119, CD146, HLA ABC showed positive correlation with CFU; female BM-MSC were more potent in T-cell proliferation suppression (Siegel et al., 2013). Despite such extensive qualification, proof of clinical relevance of such characteristics independently is still lacking. It is interesting to note that a well-chosen panel of assays may be a good indication of therapeutic activity. Three quantitative tests over different donors showed no correlation between growth rate, proliferation, cell viability and donor age/gender (Deskins et al., 2013). However, from this study it became apparent that the scores of these assays together correlated with biological activity in a wound healing assay.

References

Antwiler, G.D., Nguyen, K.T. and McNiece, I.K. (2009) Bioreactor design and implementation, in *Stem Cell Bioengineering* (eds B. Parekkadan and M. Yarmush), Artech House, Norwood, MA, pp. 49–62.

Aranda, P., Agirre, X., Ballestar, E. *et al.* (2009) Epigenetic signatures associated with different levels of differentiation potential in human stem cells. *PLoS.ONE*, **4** (11), e7809.

Aranguren, X.L., Luttun, A., Clavel, C. *et al.* (2007) *In vitro* and *in vivo* arterial differentiation of human multipotent adult progenitor cells. *Blood*, **109** (6), 263–4.

Aranguren, X.L., McCue, J.D., Hendricx, B. *et al.* (2008) Multipotent adult progenitor cells sustain function of ischemic limbs in mice. *J. Clin. Invest.*, **118** (2), 505–14.

Aranguren, X.L., Pelacho, B., Penuelas, I. *et al.* (2011) MAPC transplantation confers a more durable benefit than AC133+ cell transplantation in severe hind limb ischemia. *Cell Transplant.*, **20** (2), 259–69.

Badenes, S.M., Fernandes, T.G., Rodrigues, C.A. *et al.* (2015) Scalable expansion of human-induced pluripotent stem cells in xeno-free microcarriers. *Methods Mol. Biol.*, **1283**, 23–9.

Bersenev A. 2015. Trends in cell therapy clinical trials 2011–2014. *Cell Trials blog*: available at: http://celltrials.info/2015/02/14/trends-2014/

Boozer, S., Lehman, N., Lakshmipathy, U. *et al.* (2009) Global characterization and genomic stability of human MultiStem, a multipotent adult progenitor cell. *Journal of Stem Cells*, **4** (1), 17–28.

Bork, S., Pfister, S., Witt, H., Horn, P., Korn, B. et al. 2010. DNA methylation pattern changes upon long-term culture and aging of human mesenchymal stromal cells. *Aging Cell*, **9**(1), 54–63.

Bravery, C.A., Carmen, J., Fong, T. *et al.* (2013) Potency assay development for cellular therapy products: An ISCT review of the requirements and experiences in the industry. *Cytotherapy*, **15** (1), 9–19.

Brindley, D.A., Davie, N.L., Culme-Seymour, E.J. *et al.* (2012) Peak serum: implications of serum supply for cell therapy manufacturing. *Regen. Med.*, **7** (1), 7–13.

Busch, S.A., Hamilton, J.A., Horn, K.P. *et al.* (2011) Multipotent adult progenitor cells prevent macrophage-mediated axonal dieback and promote regrowth after spinal cord injury. *Journal of Neuroscience*, **31** (3), 944–53.

Carmelo, J.G., Fernandes-Platzgummer, A., Cabral, J.M. and da Silva, C.L. (2015) Scalable *ex vivo* expansion of human mesenchymal stem/stromal cells in microcarrier-based stirred culture systems. *Methods Mol. Biol.*, **1283**, 147–59.

Castiello, L., Sabatino, M., Zhao, Y. *et al.* (2013) Quality controls in cellular immunotherapies: rapid assessment of clinical grade dendritic cells by gene expression profiling. *Mol. Ther.*, **21** (2), 476–84.

Chase, L.G., Yang, S., Zachar, V. *et al.* (2012) Development and characterization of a clinically compliant xeno-free culture medium in good manufacturing practice for human multipotent mesenchymal stem cells. *Stem Cells Transl.Med.*, **1** (10), 750–58.

Chen, A.K., Chen, X., Choo, A.B. *et al.* (2011) Critical microcarrier properties affecting the expansion of undifferentiated human embryonic stem cells. *Stem Cell Res.*, **7** (2), 97–111.

Chen, A.K., Reuveny, S. and Oh, S.K. (2013) Application of human mesenchymal and pluripotent stem cell microcarrier cultures in cellular therapy: Achievements and future direction. *Biotechnol. Adv.*, **31** (7), 1032–46.

Chinnadurai, R., Garcia, M.A., Sakurai, Y. *et al.* (2014) Actin cytoskeletal disruption following cryopreservation alters the biodistribution of human mesenchymal stromal cells *in vivo*. *Stem Cell Reports*, **3** (1), 60–72.

Cyranoski, D. (2014) Next-generation stem cells cleared for human trial. *Nature*. http: //www.nature.com/news/next-generation-stem-cells-cleared-for-human-trial-1.15897

Dahl, J.A., Duggal, S., Coulston, N. *et al.* (2008) Genetic and epigenetic instability of human bone marrow mesenchymal stem cells expanded in autologous serum or fetal bovine serum. *Int. J. Dev. Biol.*, **52** (8), 1033–42.

Dao, M.A., Tate, C.C., Aizman, I. *et al.* (2011) Comparing the immunosuppressive potency of naive marrow stromal cells and Notch-transfected marrow stromal cells. *J. Neuroinflammation.*, **8**, 133.

Dao, M.A., Tate, C.C., McGrogan, M. and Case, C.C. (2013) Comparing the angiogenic potency of naive marrow stromal cells and Notch-transfected marrow stromal cells. *J. Transl.Med.*, **11**, 81.

Declercq, H.A., De, C.T., Krysko, O. *et al.* (2013) Bone grafts engineered from human adipose-derived stem cells in dynamic 3D-environments. *Biomaterials*, **34** (4), 1004–17.

Deskins, D.L., Bastakoty, D., Saraswati, S. *et al.* (2013) Human mesenchymal stromal cells: identifying assays to predict potency for therapeutic selection. *Stem Cells Transl. Med.*, **2** (2), 151–8.

Dominici, M., Le, B.K., Mueller, I. *et al.* (2006) Minimal criteria for defining multipotent mesenchymal stromal cells. The International Society for Cellular Therapy position statement. *Cytotherapy*, **8** (4), 315–17.

Dos Santos, F., Andrade, P.Z., Eibes, G. *et al.* (2011) *Ex vivo* expansion of human mesenchymal stem cells on microcarriers. *Methods Mol. Biol.*, **698**, 189–98.

Dos Santos, F., Campbell, A., Fernandes-Platzgummer, A. *et al.* (2014) A xenogeneic-free bioreactor system for the clinical-scale expansion of human mesenchymal stem/stromal cells. *Biotechnology and Bioengineering*, **111** (6), 1116–27.

Eggenhofer, E., Popp, F.C., Mendicino, M. *et al.* (2013) Heart grafts tolerized through third-party multipotent adult progenitor cells can be retransplanted to secondary hosts with no immunosuppression. *Stem Cells Transl.Med.*, **2** (8), 595–606.

Eyckmans, J., Lin, G.L. and Chen, C.S. (2012) Adhesive and mechanical regulation of mesenchymal stem cell differentiation in human bone marrow and periosteum-derived progenitor cells. *Biol.Open.*, **11**, 1058–68.

Fan, Y., Hsiung, M., Cheng, C. and Tzanakakis, E.S. (2014) Facile engineering of xeno-free microcarriers for the scalable cultivation of human pluripotent stem cells in stirred suspension. *Tissue Eng Part A*, **20** (3–4), 588–99.

Fierro, F.A., Kalomoiris, S., Sondergaard, C.S. and Nolta, J.A. (2011) Effects on proliferation and differentiation of multipotent bone marrow stromal cells engineered to express growth factors for combined cell and gene therapy. *Stem Cells*, **29** (11), 1727–37.

Francois, M., Copland, I.B., Yuan, S. *et al.* (2012) Cryopreserved mesenchymal stromal cells display impaired immunosuppressive properties as a result of heat-shock response and impaired interferon-gamma licensing. *Cytotherapy*, **14** (2), 147–52.

Galateanu, B., Dimonie, D., Vasile, E. *et al.* (2012) Layer-shaped alginate hydrogels enhance the biological performance of human adipose-derived stem cells. *BMC Biotechnol.*, **12**, 35.

Goh, T.K., Zhang, Z.Y., Chen, A.K. *et al.* (2013) Microcarrier culture for efficient expansion and osteogenic differentiation of human fetal mesenchymal stem cells. *Biores. Open Access.*, **2** (2), 84–97.

Hess, D.C., Sila, C.A., Furlan, A.J. *et al.* (2014) A double-blind placebo-controlled clinical evaluation of MultiStem for the treatment of ischemic stroke. *Int. J. Stroke*, **9** (3), 381–6.

Jacobs, S.A., Pinxteren, J., Roobrouck, V.D. *et al.* (2013) Human multipotent adult progenitor cells are non-immunogenic and exert potent immunomodulatory effects on alloreactive T-cell responses. *Cell Transplant.*, **22** (10), 1915–2.

Jiang, Y., Jahagirdar, B.N., Reinhardt, R.L. *et al.* (2002) Pluripotency of mesenchymal stem cells derived from adult marrow. *Nature*, **418** (6893), 41–9.

Keung, E.Z., Nelson, P.J. and Conrad, C. (2013) Concise review: Genetically engineered stem cell therapy targeting angiogenesis and tumor stroma in gastrointestinal malignancy. *Stem Cells*, **31** (2), 227–35.

Kovacsovics-Bankowski, M., Streeter, P.R., Mauch, K.A. *et al.* (2009) Clinical scale expanded adult pluripotent stem cells prevent graft-versus-host disease. *Cell Immunol.*, **255** (1–2), 55–60.

Lee, S., Kim, H.S., Roh, K.H. *et al.* (2015) DNA methyltransferase inhibition accelerates the immunomodulation and migration of human mesenchymal stem cells. *Sci. Rep.*, **5**, 8020.

Lehman, N., Cutrone, R., Raber, A. *et al.* (2012) Development of a surrogate angiogenic potency assay for clinical-grade stem cell production. *Cytotherapy.*, **14** (8), 994–1004.

Mays, R.W., Borlongan, C.V., Yasuhara, T. *et al.* (2010) Development of an allogeneic adherent stem cell therapy for treatment of ischemic stroke. *J. Exp. Stroke Transl. Med.*, **3** (1), 34–46.

Maziarz, R.T., Devos, T., Bachier, C. *et al.* (2012) Prophylaxis of acute GVHD using MultiStem stormal cell therapy: Preliminary results after administration of single or multiple doses in a Phase 1 trial. *Biol. Blood Marrow Transplant.*, **18** (2), S264–5.

McBeath, R., Pirone, D.M., Nelson, C.M. *et al.* (2004) Cell shape, cytoskeletal tension, and RhoA regulate stem cell lineage commitment. *Dev. Cell*, **6** (4), 483–95.

Moll, G., Alm, J.J., Davies, L.C. *et al.* (2014) Do cryopreserved mesenchymal stromal cells display impaired immunomodulatory and therapeutic properties? *Stem Cells*, **32** (9), 2430–42.

Mora-Lee, S., Sirerol-Piquer, M.S., Gutierrez-Perez, M. *et al.* (2012) Therapeutic effects of hMAPC and hMSC transplantation after stroke in mice. *PLoS.ON*, **7** (8), e43683.

Oskowitz, A., McFerrin, H., Gutschow, M. *et al.* (2011) Serum-deprived human multipotent mesenchymal stromal cells (MSCs) are highly angiogenic. *Stem Cell Res.*, **6** (3), 215–25.

Otsuru, S., Hofmann, T.J., Raman, P. *et al.* (2015) Genomic and functional comparison of mesenchymal stromal cells prepared using two isolation methods. *Cytotherapy*, **17** (3), 262–70.

Pattasseril, J., Varadaraju, H., Lock, L. and Rowley, J.A. (2013) Downstream technology landscape for large-scale therapeutic cell processing. *BioProcess International*, **11** (3), 46–52.

Penn, M.S., Ellis, S., Gandhi, S. *et al.* (2012) Adventitial delivery of an allogeneic bone marrow-derived adherent stem cell in acute myocardial infarction: Phase I clinical study. *Circulation Research*, **110** (2), 304–11.

Pritchett, T. and Little, L. (2012) Hard cell: potency testing for cellular therapies. *BioProcess International*, **10**, 36–48.

Reading, J.L., Yang, J.H., Sabbah, S. *et al.* (2013) Clinical-grade multipotent adult progenitor cells durably control pathogenic T cell responses in human models of transplantation and autoimmunity. *Journal of Immunology*, **190** (9), 4542–52.

Roobrouck, V.D., Clavel, C., Jacobs, S.A. *et al.* (2011) Differentiation potential of human postnatal mesenchymal stem cells, mesoangioblasts, and multipotent adult progenitor cells reflected in their transcriptome and partially influenced by the culture conditions. *Stem Cells*, **29** (5), 871–82.

Rowley, J.A., Abraham, E., Campbell, A. and Brandwein, H. (2012) Meeting lot-size challenges of manufacturing adherent cells for therapy. *BioProcess International*, **10** (3), 16–22.

Ryu, J.C., Yue, Q., Carr, C. *et al.* (2011) Molecular Imaging of the pro-angiogenic paracrine effects of stem cell therapy. *Circulation*, **124**, A9926.

Santos, F., Andrade, P.Z., Abecasis, M.M. *et al.* (2011) Toward a clinical-grade expansion of mesenchymal stem cells from human sources: a microcarrier-based culture system under xeno-free conditions. *Tissue Eng. Part C Methods*, **17** (12), 1201–10.

Siddappa, R., Licht, R., van, B.C. and de, B.J. (2007) Donor variation and loss of multipotency during *in vitro* expansion of human mesenchymal stem cells for bone tissue engineering. *J. Orthop. Res.*, **25** (8), 1029–41.

Siegel, G., Kluba, T., Hermanutz-Klein, U. *et al.* (2013) Phenotype, donor age and gender affect function of human bone marrow-derived mesenchymal stromal cells. *BMC Med.*, **11**, 146.

Teklemariam, T., Purandare, B., Zhao, L. and Hantash, B.M. (2014) Inhibition of DNA methylation enhances HLA-G expression in human mesenchymal stem cells. *Biochemical and Biophysical Research Communications*, **452** (3), 753–9.

Trojahn Kolle, S.F., Oliveri, R.S., Glovinski, P.V. *et al.* (2013) Pooled human platelet lysate versus fetal bovine serum-investigating the proliferation rate, chromosome stability and angiogenic potential of human adipose tissue-derived stem cells intended for clinical use. *Cytotherapy*, **15** (9), 1086–97.

Tseng, P.C., Young, T.H., Wang, T.M. *et al.* (2012) Spontaneous osteogenesis of MSCs cultured on 3D microcarriers through alteration of cytoskeletal tension. *Biomaterials*, **33** (2), 556–64.

van't Hof, W., Mal, N., Huang, Y. *et al.* (2007) Direct delivery of syngeneic and allogeneic large-scale expanded multipotent adult progenitor cells improves cardiac function after myocardial infarct. *Cytotherapy*, **9** (5), 477–87.

Walker, P.A., Shah, S.K., Jimenez, F. *et al.* (2010) Intravenous multipotent adult progenitor cell therapy for traumatic brain injury: preserving the blood brain barrier via an interaction with splenocytes. *Exp. Neurol.*, **225** (2), 341–52.

Walker, P.A., Bedi, S., Shah, S.K. *et al.* (2012) Intravenous multipotent adult progenitor cell therapy after traumatic brain injury: modulation of the resident microglia population. *J. Neuroinflammation.*, **9** (1), 228.

Zeiger, A.S., Loe, F.C., Li, R. *et al.* (2012) Macromolecular crowding directs extracellular matrix organization and mesenchymal stem cell behavior. *PLoS.ONE*, **7** (5), e37904.

Zeisberger, S.M., Zoller, S., Riegel, M. *et al.* (2010) Optimization of the culturing conditions of human umbilical cord blood-derived endothelial colony-forming cells under xeno-free conditions applying a transcriptomic approach. *Genes Cells*, **15** (7), 671–87.

Zhang, D. and Kilian, K.A. (2013) The effect of mesenchymal stem cell shape on the maintenance of multipotency. *Biomaterials*, **34** (16), 3962–9.

Zhu, Y., Song, X., Han, F., Li, Y., Wei, J. and Liu, X. (2015a) Alteration of histone acetylation pattern during long-term serum-free culture conditions of human fetal placental mesenchymal stem cells. *PLoS. ONE*, **10**(2), e0117068.

Zhu, Y., Song, X., Wang, J. *et al.* (2015b) Placental mesenchymal stem cells of fetal origin deposit epigenetic alterations during long-term culture under serum-free condition. *Expert. Opin. Biol. Ther.*, **15** (2), 163–80.

4

The Scale-up of Human Mesenchymal Stem Cell Expansion and Recovery

Thomas R. J. Heathman[1], *Qasim A. Rafiq*[1,3], *Karen Coopman*[1], *Alvin W. Nienow*[1,2] *and Christopher J. Hewitt*[1,3]

[1] *Centre for Biological Engineering, Loughborough University, Leicestershire, UK*
[2] *Centre for Bioprocess Engineering, University of Birmingham, UK*
[3] *Aston Medical Research Institue, School of Life and Health Sciences, Aston University, Birmingham, UK*

4.1 Introduction

The last 25 to 30 years has seen a healthcare revolution based on the production of human proteins in genetically-modified cells, with an estimated current market value of approximately £93 billion a year (Want et al., 2012). There is now an opportunity to replicate this success in a new industry using cell therapies, based on various types of stem cells. Such treatments may constitute the next healthcare revolution, creating biological therapies that replace or restore normal biological function lost through failure and/or disease. Science and medicine has revealed the potential of cell therapies; however, for this to be realised, scalable and robust manufacturing processes must be developed for the reproducible production of clinically-relevant numbers of fully functional stem cells (Rafiq et al., 2013b).

It is well established that for all products made either by cell culture or fermentation, the process must be defined with equally defined critical quality attributes (CQAs) that can be monitored and controlled to ensure product safety as well as efficacy (Rafiq and Hewitt, 2015). Even as far back as Babylonian times (ca. 1730 BC) the role of the head brewer when manufacturing beer was seen as an important one (Hewitt 2003). Indeed, so important a role was it that the

Bioprocessing for Cell Based Therapies, First Edition. Edited by Che J. Connon.

punishment for a brewer guilty of malpractice was to be drowned in his or her own beer if deemed undrinkable, highlighting the importance of CQAs even then. Sir Alexander Fleming, on discovering penicillin in 1928, had no real idea of its significance and it was not until a team made up of Florey, Chain and Heatley (Lax, 2005) devised a process for the mass production of the drug that the full power of penicillin for treating infection was realised. Within the biopharmaceutical industry, sophisticated submerged fed-batch reactors are currently used for producing many important biologically active drugs (e.g. monoclonal antibodies) and their corresponding processes are regulated and approved by powerful national agencies (FDA, MHRA, etc.).

However, up to now we have been using the cell as a "factory" and it is rare that the cell forms the basis of the functional product. When developing the process to produce Herceptin (Roche, Switzerland), for example, it is irrelevant what happens to the Chinese Hamster Ovary (CHO) cell, provided it produces greater amounts of functional protein. Indeed, during protein purification in subsequent downstream processing, the cell is discarded very early on. The wild-type Chinese Hamster has 22 chromosomes; however, karyotypic analysis of CHO production cell lines found many aberrations had accumulated over time, with chromosomal duplications and deletions often corresponding to markers associated with adaptation to long-term culture (Wurm, 2013). Cell therapies represent a paradigm shift, because the cell forms the basis of the functional product and must remain the same from donor isolation through expansion, recovery, formulation and final delivery to patient. This mode of usage means the phenotype, genotype and even karyotype, on occasion, need to be shown to be the same throughout (Want et al., 2012). Unlike traditional biopharmaceutical production, stem cells cannot be immortalised in any way, as this may cause teratomas or cancers to form in the patient and is likely to remove the functionality of the product. It takes only one mortality to occur as a result of cell therapy-induced teratoma production for the field to be thrown into absolute crisis. Furthermore, stem cells need to be attached to a surface and it is unlikely they can be adapted to suspension culture whilst retaining their CQAs. All these differences present a challenge for the bioprocessing community when trying to bring cell therapies to the worldwide marketplace.

4.2 Scale-up or Scale-out

So the question for the emerging cell therapy industry is "should the manufacturing process be scaled-up or scaled-out?" (Table 4.1). For autologous cell therapies (where the donor and recipient are the same individual – so patient lots must be physically separate) a scale-out approach is more likely to be cost-effective, where automated monolayer culture platforms have already proved successful (Thomas et al., 2007; Want et al., 2012). However, the challenge of manipulating and processing multiple T-flasks or cell factories in a closed, sterile manner is not trivial. The usual process steps, routinely carried out by a trained operator, such as passaging cells, performing medium exchanges and sampling cell culture medium, require significant investment in automated manufacturing technologies, which in turn involves significant capital expenditure and will require comparability studies between both the manual and automated processes before being accepted by the regulator (Rafiq and Hewitt, 2015). In practice, this involves designing and programming robotic devices to mimic the actions of a human operator. However, defining what a human does in the laboratory is not always easy and requires detailed process mapping, but once complete, robotic platforms have already been shown to outperform a human both in quantity of T-flask handled (~10x more) and in the consistency of the bioprocessing output by reducing process variation (Thomas et al, 2007, 2009).

For allogeneic cell therapy processes (where the donor and recipient are different individuals), scale-up approaches are more likely to be employed. However, the challenges are even more significant from a cell expansion perspective. The use of stirred-tank bioreactors for the large-scale manufacture of biopharmaceuticals is now relatively

Table 4.1 A risk/reward and cost comparison of the scale-out (automation) or scale-up (bioreactor) options.

Automated Platform	Bioreactor Platform
1) Transfer of existent manual process	1) Transformation of process
2) Favoured by the Regulators	2) Not favoured by the Regulator
3) High cost but low risk	3) Low cost but high risk
4) High value	4) High potential financial reward
5) Shorter time to market/clinic	5) Longer time to market/clinic

routine (Nienow, 2006). However, given that the majority of cell therapy candidates employ anchorage-dependent cells, microcarriers are required for stirred-tank bioreactor expansion (Croughan et al., 1987; Ibrahim and Nienow, 2004). Moving from a manual, T-flask based process to a microcarrier stirred-tank bioreactor system is a substantial transition, which requires significant process and product understanding. Furthermore, harvesting the cells from the microcarriers, only considered recently, whilst maintaining the cells' CQAs is different from monolayer culture and poses a significant engineering challenge (Nienow et al., 2014). As well as the physical challenge of removing the cells from the microcarriers, there remains a challenge in demonstrating comparability of cell CQAs with the monolayer process, which will be essential to effectively scale up the process.

Unfortunately, within the biopharmaceutical industry, scale-up is usually the final step in any "Research and Development programme", leading to the large-scale production of recombinant or natural proteins by cell culture in stirred-tank vessels (Rafiq and Hewitt, 2015). It is fair to say that most basic researchers do not have a defined process nor appreciate the challenges associated with translation, and often (although this is getting better) most engineers do not understand the biology. However, it is certainly the case that engineers do not get involved early enough and that there is a "throw it over the wall" mentality with very little in-depth communication between the two groups of people (Rafiq and Hewitt, 2015). For biologists it is important to understand that the process of scaling-up cell culture is often governed by a number of important engineering considerations, including physical aspects of the agitation/agitator strategy (Table 4.2: Hewitt and Nienow, 2007). Importantly, these are all generally system specific. Clearly, it is not simply a matter of increasing culture and vessel volume nor simply "making the pot bigger". Therefore, it can be surprising for some when there is a detrimental change in the bioprocessing outcomes as you move up through the scales. Indeed it is often observed that the biomass yield and any growth-associated products often decrease significantly on the scale-up of an aerobic process, and this should be considered early on in the development phase (Hewitt and Nienow, 2007). These detrimental changes may arise when one moves from the unaerated and unagitated T-flask to the agitated and gassed environment of a stirred tank bioreactor and then up through the scales (Hewitt and Solomons, 1996; Rafiq and Hewitt, 2015).

Table 4.2 Physical aspects of the agitation/agitator requiring consideration and biological aspects that are often system specific (Hewitt and Nienow, 2007).

Physical aspects	Biological aspects
1) Mass transfer performance	1) Growth and productivity
2) Heat transfer	2) Nutrient and other additive requirements including oxygen
3) Unaerated power draw (or mean specific energy dissipation rate, $\overline{\varepsilon}_T$ W/kg).	3) CO_2 evolution and Respiratory Quotient
4) Aerated power draw (or aerated ($\overline{\varepsilon}_T$)W/kg)	4) Sensitivity to O_2 and CO_2 concentration.
5) Flow close to the agitator-single phase and air-liquid	5) pH range and sensitivity
6) Variation in local specific energy dissipation rates, $\overline{\varepsilon}_T$ W/kg	6) Operating temperature range
7) Air dispersion capability	7) "Shear sensitivity"
8) Bulk fluid- and air-phase mixing	

When comparing a T-flask with a continuous stirred tank bioreactor, it becomes obvious that the respective engineering environments are very different, and it should therefore be much less surprising if there are changes in bioprocess performance. For a T-flask gas (CO_2/O_2) exchange is passive, relying solely on diffusion; there is crude pH control *via* a visual inspection of a bicarbonate buffering system and temperature control is through placement in a controlled incubator. For a bioreactor, pH control is by automatic titration, temperature control is very tight and gas exchange is active *via* mechanical agitation and aeration where dissolved oxygen (dO_2) is controllable. Reference to the damaging effects on cells of fluid mechanical stress or so-called "shear damage", associated with the introduction of mechanical agitation and aeration, is frequently made to predict poor process performance when moving from the unagitated and unsparged conditions of a T-flask (Hewitt et al., 1998; Boswell et al., 2003).

However, for all single-cell types used for producing bioactive molecules, the effects of fluid mechanical stress from mechanical agitation have largely been discounted during the normal operating range, even for animal cells that were thought to be more susceptible to such damage because of a lack of cell wall (Nienow, 2006, 2014). Indeed, any damage to freely suspended animal cells is more likely to be due to bubbles bursting at the medium surface (Oh et al., 1989, 1992; Nienow, 2006; Thomas, 1990). Fluid dynamic modelling of bursting bubbles has shown that there are huge local specific energy dissipation rates (10^4–10^5 W/kg) associated with it, much higher than that associated with mechanical agitation. However, it is again recognised that the addition of the surfactant Pluronic F68 to the culture medium essentially eliminates damage and it does so by preventing bubble/cell attachment and reduces medium surface tension so that the animal cells do not experience such high local stresses (Thomas, 1990). It is important to emphasise here that fluid mechanical stresses have been shown to have no effect only on the intended bioprocessing outcomes, which in all the cases studied so far have been the optimal production of intra and extra cellular products rather than the cells themselves. Therefore it remains to be seen whether such forces will affect the CQAs of stem cells. Furthermore, a major perceived problem is that where microcarriers are used, cells may be removed from the surfaces by mechanical agitation or by microcarrier to microcarrier collision (Hewitt et al., 2006). This issue will be more fully addressed later. There is also concern that such collisions may also damage the microcarriers themselves, potentially leaving small particulates of microcarrier material that will need to be removed from the product before delivery to the patient.

It has been shown that the most likely cause of observable poor performance arising as the stirred tank bioreactor process is scaled, is because of the measureable presence of spatial and temporal concentration gradients due to longer mixing times and poorer mixing generally in comparison to the small scale, particularly when a fed-batch strategy (currently preferred by industry) is employed with pH control by automatic titration (Enfors et al., 2001; Hewitt and Nienow, 2007). For cell therapies, where cell quality is of paramount importance, any fluctuations in the culture environment may have an impact on the quality and consistency of the cells, since it is highly probable that temporal and spatial gradients will indeed exist since a perfusion strategy is the likely way forward (Rafiq and Hewitt, 2015). Therefore the effects of cells circulating through such gradients on their quantity, quality and viability will need to be quantified and mechanisms of mitigating against these effects, if any, identified.

4.3 Understanding the Small Scale

Currently, the routine expansion of hMSCs by biologists typically involves the *in vitro* culture of cells as a monolayer in T-flasks under static conditions, with any measurements usually limited to cell viability, morphology, confluency and those related to post-culture functionality. However, it is our contention that in order to inform the basis of the development of the larger-scale production of hMSCs (e.g. for the accurate calculation of nutrient feeds for perfusion), the factors controlling the process need to be fully understood by the engineer. A key analytical tool used in engineering is the mass balance (Rafiq et al., 2013c). However, to date, it has not been possible to develop an accurate mass balance calculation for hMSC T-flask culture, because nutrient usage, metabolite production and headspace gas composition are not routinely measured. Clearly, the measurement and potential for control of culture parameters is crucial, because even small changes in culture conditions can have a significant effect. The level of dissolved oxygen (dO_2) in the growth medium is also extremely important and as such should always be measured and, where possible, controlled. As identified previously (Rafiq et al., 2013c), unfortunately there is a disparity between the nomenclature used to describe the oxygen concentration present in the cells' growth medium in the bioprocess engineering and life sciences fields, which may lead to confusion when trying to understand the significance of results.

With so much experience already available concerning the growth of animal cells on the large scale, it would seem advantageous to bring the two into line. The levels of oxygen usually described by the term "normoxia" (20–21% v/v oxygen in the headspace) is equivalent to 100% dO_2 in the culture medium when saturated with respect to air; whilst the term "hypoxia" (2–5% v/v oxygen) is equivalent to 10–25% dO_2. The problem is further exacerbated because when compared to the dO_2 that hMSCs experience in the body (discussed later), 100% dO_2 ("normoxia") would generally be considered "hyperoxic" physiologically, whilst a reduced dO_2 is more likely to be "normoxic" physiologically. Therefore, by considering the oxygen concentration in terms of the nominal % dO_2 with respect to saturation in the liquid medium, which is the standard nomenclature in biochemical engineering, some of these ambiguities can be reduced and such data will be easier to compare.

Recently, for the first time (Rafiq et al., 2013c), we provided a quantitative analysis of the changes in concentration of glucose, lactate and ammonium with time during a standard hMSC monolayer culture, used by Lonza, over 4 passages, under 100% and 20% dissolved oxygen (dO_2), where either a 100%, 50% or 0% growth medium exchange was

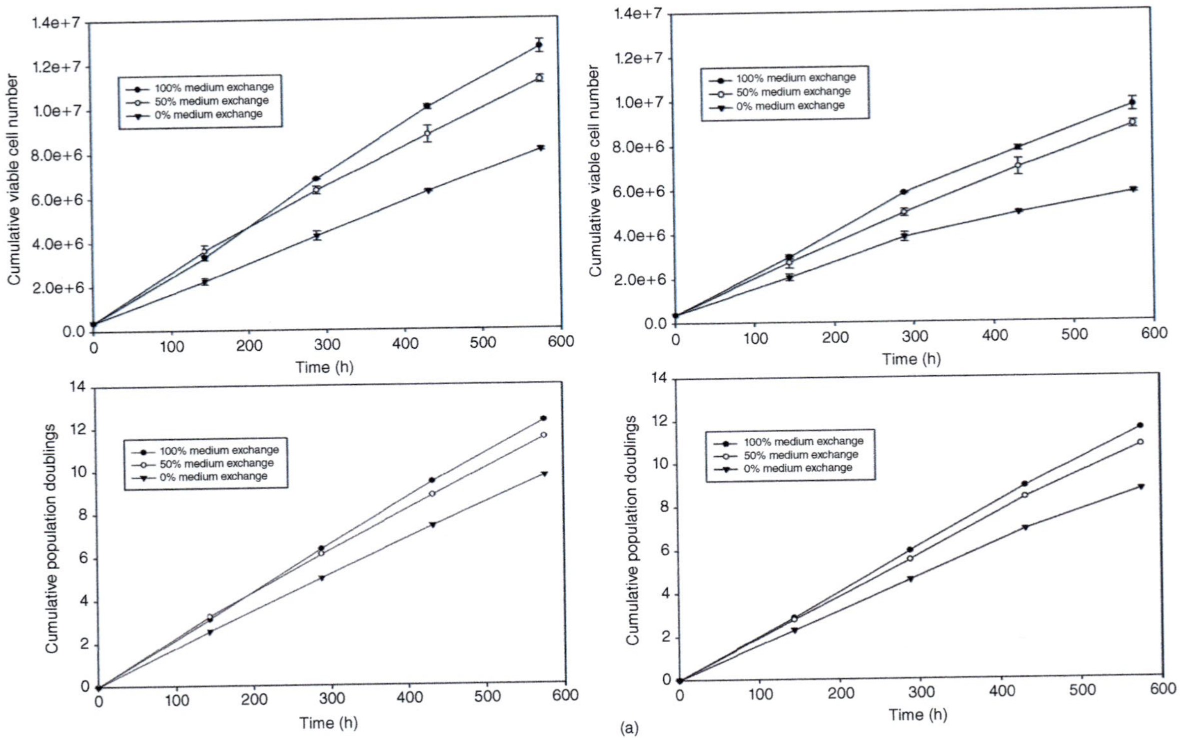

(a)

Figure 4.1 (a) Cumulative cell numbers (upper) and population doublings (lower) at 100% dO_2 at saturation (20% O_2 v/v gas phase) (left) and 20% dO_2 at saturation (4% O_2 v/v gas phase) (right) with 100%, 50% and 0% medium exchange (mean value ± SD, n = 4); (b) The effect of medium exchange and oxygen concentration on metabolite consumption and production (mean value ± SD, n = 4) (Rafiq et al., 2013c).

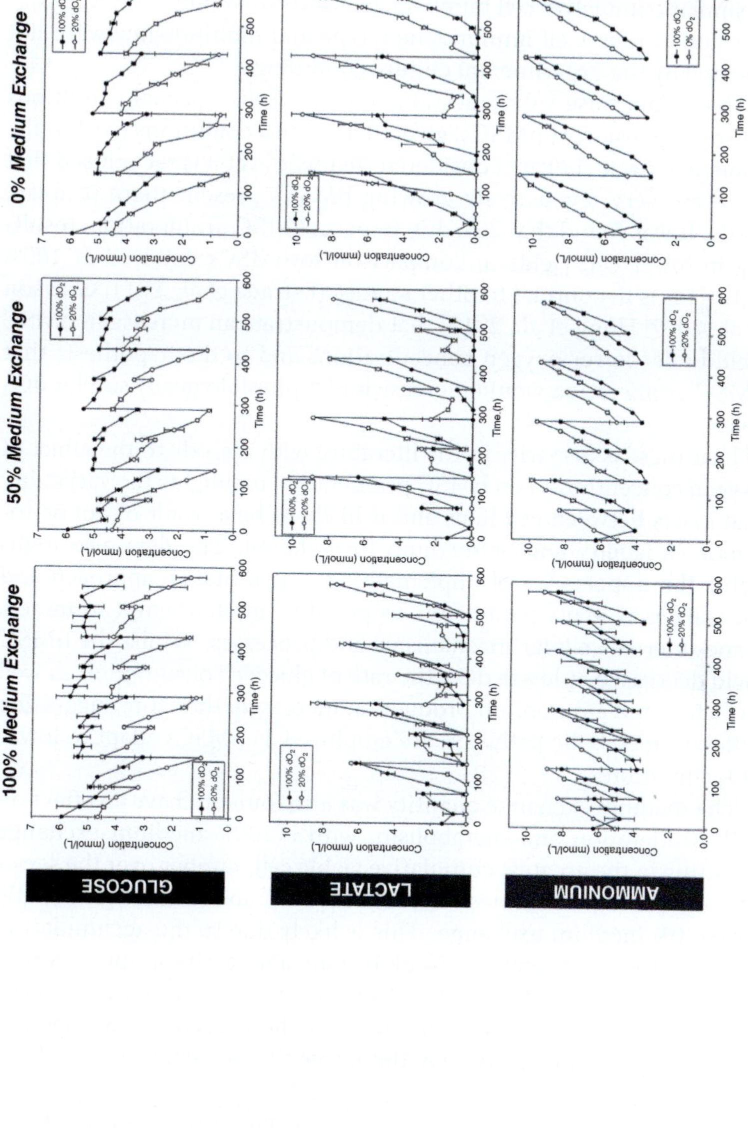

Figure 4.1 (Continued)

performed after 72 h in culture (Figure 4.1). Yield coefficients, specific growth rates (h^{-1}) and doubling times (h) were calculated in all cases (Table 4.3). The 100% dO_2 flasks outperformed the 20% dO_2 flasks with respect to cumulative cell number, with the latter consuming more glucose and producing more lactate and ammonium. Furthermore, the 100% and 50% medium exchange conditions resulted in similar cumulative cell numbers, whilst the 0% conditions were significantly lower. Cell immunophenotype and multipotency were not affected by the experimental culture conditions.

These data raise valid questions regarding the optimal conditions for the expansion of hMSCs, given that *in vivo* conditions (within the bone marrow niche) are considered "hypoxic", yet it is recognised that there are very few actively growing hMSCs present there (Caplan, 2007). It was found that 20% dO_2 impacts hMSC proliferation, resulting in lower cell yields in comparison to hMSCs cultured at 100% dO_2. This is in contrast to other studies (Estrada et al., 2011; Grayson et al., 2007; Hung et al., 2012) that demonstrate an increase in hMSC yield under lower oxygen concentrations and to the hypothesis that hMSC proliferation would increase under physiologically similar dO_2 levels.

That there is disparity in the literature with regards to the effect of oxygen concentration on hMSC proliferation highlights the variability that exists between cell lines and is likely to be a result of donor-to-donor variability and/or medium formulation. The data also highlights the importance of implementing a quantitative approach and measuring as many parameters as possible, in an attempt to understand internal cellular mechanisms and processes. Whilst the hMSC yield decreased at lower dO_2, the rate of glucose consumption as well as lactate and ammonium production increased, therefore suggesting different metabolic pathways are employed as hMSCs adapt to lower dO_2 conditions.

The medium exchange quantity was also found to have an effect on cell proliferation and metabolism, with a 100% medium exchange resulting in the greatest cumulative viable cell number over the series of four passages, followed closely by 50% and then, significantly lower, 0% medium exchange. This is likely due to the accumulation of inhibitory concentrations of lactate and ammonium or other unknown waste products. There was a clear correlation between the concentration of these metabolites and the percentage of medium exchange performed, whereby the lower the percentage of medium

Table 4.3 (a) Impact of oxygen concentration and medium exchange on specific growth rate, fold increase and doubling time. 100% dO_2 values are emboldened and underlined, whereas 20% dO_2 values are not; (b) Impact of dissolved oxygen concentration and medium exchange on specific glucose consumption, lactate production and yield of lactate on glucose. Negative values indicate consumption. 100% dO_2 values are emboldened and underlined, whereas 20% dO_2 values are not (Rafiq et al. 2013c).

(a)	100% Medium Exchange				50% Medium Exchange				0% Medium Exchange			
Passage Number	1	2	3	4	1	2	3	4	1	2	3	4
Specific growth rate (h^{-1})	0.015	0.016	0.015	0.014	0.016	0.014	0.013	0.013	0.012	0.012	0.012	0.011
	0.014	0.015	0.014	0.013	0.014	0.013	0.014	0.012	0.011	0.011	0.011	0.009
Fold increase	8.8	9.4	8.5	7.1	9.7	7.2	6.6	6.5	6.0	5.4	5.3	4.9
	7.4	8.4	7.8	6.3	7.0	6.7	7.2	5.2	5.0	5.0	4.9	3.5
Doubling Time (h)	45.8	44.5	46.6	50.8	44.0	50.6	53.1	53.4	55.9	58.9	60.0	62.6
	49.9	47.0	48.6	54.4	51.3	52.6	50.6	60.3	62.3	62.3	62.6	79.3

(Continued)

Table 4.3 (*Continued*)

(b)	100% Medium Exchange				50% Medium Exchange				0% Medium Exchange			
Passage Number	1	2	3	4	1	2	3	4	1	2	3	4
Specific glucose consumption rate (pmol cell^{-1} day^{-1})	−10.55 −16.60	−6.34 −13.62	−5.89 −12.40	−4.55 −12.38	−10.41 −16.22	−6.58 −14.45	−8.14 −14.15	−7.16 −15.19	−12.22 −18.35	−7.42 −16.35	−7.44 −15.68	−5.90 −20.04
Specific lactate production rate (pmol cell^{-1} day^{-1})	21.96 22.39	22.89 35.26	20.22 25.17	23.76 37.67	20.18 32.23	23.19 41.90	21.30 26.93	23.37 40.03	19.33 41.44	23.55 43.93	23.04 33.37	19.32 46.24
Yield of lactate from glucose (mol mol^{-1})	2.08 1.35	3.61 2.59	3.43 2.03	5.22 3.04	1.94 1.99	3.52 2.90	2.62 1.90	3.26 2.63	1.58 2.26	3.17 2.69	3.10 2.13	3.28 2.31

exchange performed, the higher the concentration of both ammonium and lactate at the end of each passage. This demonstrates that ≥50% medium exchange would aid cell proliferation, highlighting potential cost savings over performing a 100% medium exchange, perhaps proving significant for large-scale bioprocessing. Importantly, despite the variation in experimental conditions, there was no discernible effect on cell quality, as the cells retained both their immunophenotype and multipotency throughout. Furthermore, this work highlights the likely key process engineering parameters for subsequent larger bioprocess development (Rafiq et al., 2013c), which is the same approach taken by the pharmaceutical industry more than two decades ago to develop large-scale mammalian cell fermentation for the manufacture of therapeutic recombinant proteins (Jones et al., 2012; Mason and Hoare, 2006).

4.4 Microcarrier Screening

As mentioned previously, human mesenchymal stem cells need to be attached to a surface to grow and cannot be adapted to suspension culture. Microcarriers are particles of various shapes and sizes (100 μm–2 mm) that when in supsension in a suitable growth medium can support high cell densities by providing a large surface area for anchorage dependent growth (Table 4.4; Hewitt et al., 2006; Ibrahim and Nienow, 2004). Microcarriers are also able to maintain a homogenous physical environment and provide relatively easy separation of cells from the growth medium at the end of the process. However, a major problem is the perception that cells can be accidently removed from the microcarrier surfaces relatively easily by agitation in a bioreactor. Were this to occur, removal would likely be accompanied by a loss in viable and functional cells – not ideal given that the cell forms the basis of the product. For conventional animal cell culture, macroporous microcarriers provide some protection to the cells, since cells can penetrate and grow in the internal spaces. However, since single cells need to be recovered from the particle at the end of the culture process, it is unlikely macroporous microcarriers will be employed here.

Agitation intensities sufficient for microcarrier suspension in bioreactors may be kept to energy dissiptation rates below those at which damage occurs, by the use of a down-pumping hydrofoil impeller coupled

Table 4.4 Properties of commercially available microcarriers.

Microcarrier	Manufacturer	Diameter (μm)	Matrix	Average density	Surface coating	Surface charge	Carrier porosity	Usage in this study
Mammalian protein-coated microcarriers								
Collagen	SoloHill Eng. Inc.	125–212	Polystyrene	1.02	Type I porcine collagen	None	Non-porous	√ (Coll)
Cultispher-G®	Percell-Biolytica	130–380	Type I porcine gelatin	1.04	None	None	Macroporous (porosity: 50 % pore size: 10–30 μm)	√ (CG)
Cytodex 3™	GE Healthcare	141–211	Dextran	1.04	Type I porcine collagen	None	Non-porous	√ (Cyto3)
FACT III	SoloHill Eng. Inc.	125–212	Polystyrene	1.02	Cationic Type I porcine collagen	None	Non-porous	√ (FACT)
SphereCol®	Advanced BioMatrix	125–212	Polystyrene	1.03	Type I human collagen (VitroCol®	None	Non-porous	×
Recombinant protein-coated microcarriers								
Pro-Nectin® F	Solo Hill Eng. Inc.	125–212	Polystyrene	1.02	Recombinant fibronectin	None	Non-porous	√ (Pro-F)
Xeno-free microcarriers								
Cytodex 1™	GE Healthcare	147–248	Dextran	1.03	DEAE	+	Non-porous	√ (Cyto1)

Cytopore 1 and 2™	GE Healthcare	200–280	Cotton cellulose	1.03	DEAE	+	Micro/ Macroporous (porosity: > 90% pore size: 30 μm)	×
Enhanced Attachment	Corning	125–212	Polystyrene	1.02	CellBIND®	None	Non-porous	√ (EA)
Glass	SoloHill Eng. Inc.	125–212	Polystyrene	1.02	High silica glass	None	Non-porous	×
Hillex® CT	SoloHill Eng. Inc.	90–212	Polystyrene	1.12	Cationic trimethyl ammonium	+	Non-porous	×
Hillex®	SoloHill Eng. Inc.	160–180	Dextran	1.11	Cationic trimethyl ammonium	+	Non-porous	√ (Hillex)
MicroHex™	Nunc	Side length: 125 μm Thickness: 25 μm	Polystyrene	1.05	Nunclon™surface	Not specified	Non-porous	√ (MHex)
Plastic	SoloHill Eng. Inc.	125–212	Polystyrene	1.02	None	None	Non-porous	√ (Plas)
Plastic Plus	SoloHill Eng. Inc.	125–212	Polystyrene	1.02	None	+	Non-porous	√ (Pplus)
PVA	Loughborough University	100–220	PVA	1.03	None	None	Non-porous	√ (PVA)
Synthemax II®	Corning	125–212	Polystyrene	1.02	Synthemax II®	None	Non-porous	√ (Sy)

with a contoured bioreactor base (Ibrahim and Nienow, 2004). Also by selecting a suitable medium feeding strategy, a well-mixed physical and chemical environment within the bioreactor may be achieved. However, higher levels of agitation may be required if oxygenation is *via* surface aeration alone, especially at larger scales of operation or if high cell densities are required. This may of course be complicated if the sparging of air directly into the medium becomes necessary, since cells may be damaged by air bubbles bursting at the medium surface and it is not clear if the addition of Pluronic F68 (as for animal cell culture), in mitigation, will affect the process or final product in any way, and if so, whether its presence will be permitted by the regulator.

Such microcarrier-based systems are being increasingly investigated for large-scale production of anchorage-dependent hMSCs and the selection of an appropriate microcarrier is a fundamental component of microcarrier/stirred-tank bioreactor cell expansion (Hewitt et al., 2011; Rafiq et al., 2016; Want et al., 2012). From a cell therapy processing perspective, it is expected that once selected, the microcarrier of choice will need to remain the same throughout, particularly during the later stages of clinical development (Phase II and Phase III), at which point the process is "locked in". Any change at this point is likely to constitute a major change in the manufacturing process, requiring significant comparability testing and repeats of clinical trials (Rafiq et al., 2016). Therefore, it is of paramount importance that the optimal microcarrier is selected from the outset based on a stringent selection methodology. However, there is no robust, rigorous, transferable methodology for microcarrier selection, with existing studies providing little or no reason as to why a specific microcarrier was selected.

Recently, we systematically evaluated 13 microcarriers for human bone marrow-derived MSC expansion from three donors (hMSC1 – 3), with the aim of establishing a reproducible and transferable methodology for microcarrier selection (Figure 4.2; Rafiq et al., 2016). Monolayer studies in T-flasks demonstrated input cell line variability with respect to growth kinetics and metabolite flux. hMSC1 underwent more cumulative population doublings over three passages in comparison to hMSC2 and hMSC3. Microcarrier screening studies were performed in microwell plates and 100 mL agitated spinner flasks with all three donor hMSC1 – 3. Irrespective of donor, agitated conditions were always significantly better than static conditions and relative microcarrier performance was identical, with the same microcarriers outperforming others with respect to growth kinetics and

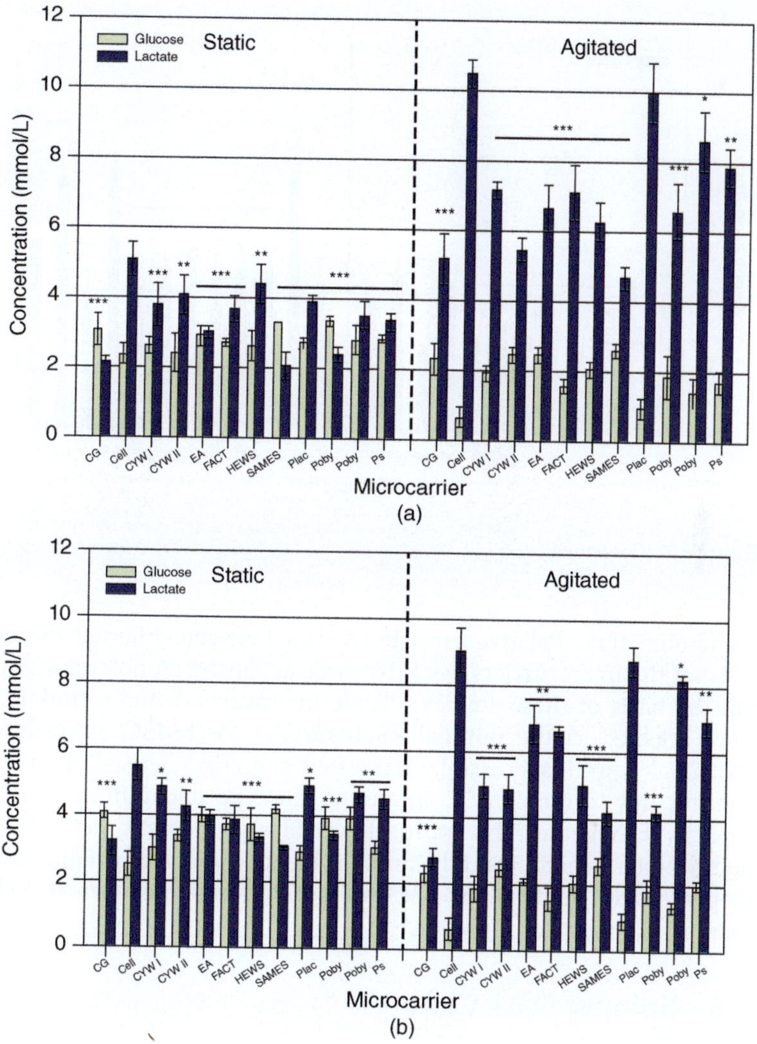

Figure 4.2 Comparison of static and agitated conditions for hMSC microcarrier culture, and a comparison of different microcarriers in 100 mL spinner flasks. Glucose consumption and lactate production for hMSC1 (a), hMSC2 (b) and hMSC3 (c). Data is presented as mean ± SD (n = 4). Significant differences in values were noted with $p < 0.05$ (*), $p < 0.01$ (**) and $p < 0.005$ (***) in comparison to the highest respective value.

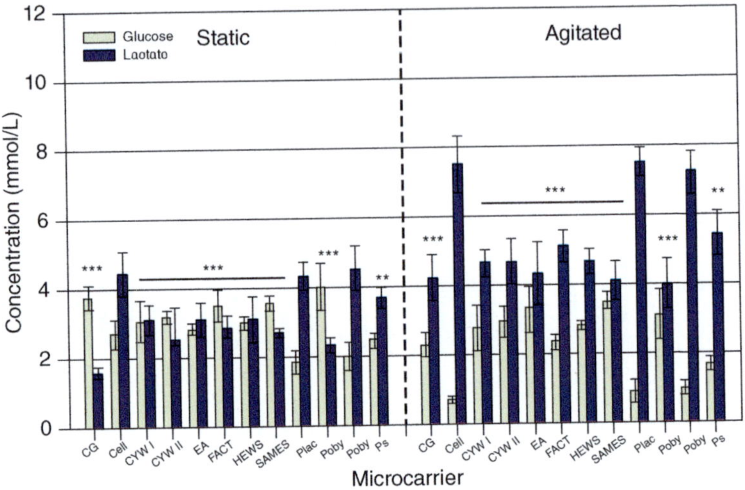

(c)

Figure 4.2 *(Continued)*

metabolite flux. Relative growth kinetics between different donor cells on the microcarriers were the same as for the monolayer study. On the basis of these studies, Plastic microcarriers and Cytodex 3 were selected as the optimal microcarriers for hMSC expansion. hMSC1 – 3 were successfully harvested and characterised, demonstrating the correct hMSC immunophenotype and differentiation capacity throughout. This approach provides a systematic method for microcarrier selection and the findings identify potentially significant bioprocessing implications for microcarrier-based allogeneic cell therapy manufacture.

4.5 Spinner Flask Culture

Clearly, from this work, certain types of microcarriers are appropriate for the cultivation of human mesenchymal stem cells and this has been shown by others (Abranches et al., 2007; Frauenschuh et al., 2007; Oh et al., 2009); however, little detailed work has been done to optimise cell microcarrier attachment protocols or cell proliferation and growth. For a microcarrier's entire surface to be available for cell attachment and growth, it needs to be fully suspended in the growth medium. This condition can be achieved by mechanical agitation (Nienow, 1997a), which also homogenises the environment with respect to substrate composition and temperature, whilst facilitating oxygen transfer to the cells and, as importantly, CO_2 stripping from the medium (Nienow, 2006). However, even for animal cells, little work has been done to study the impact of the fluid dynamic conditions on bioreactor performance with microcarriers, because early in the development of that technology, free suspension culture became available and quickly dominated because it is easier to perform (Kumar et al., 2008; Nienow, 2006).

Increased agitation might cause damage to the microcarriers or cells, as found under some conditions for human fibroblasts (Croughan et al., 1988, 1989) and bovine embryonic kidney cells (Cherry and Papoutsakis, 1986). None of these issues have been thoroughly studied or resolved for microcarrier cultures of hMSCs. Most literature addressing damage to particles of a similar size to microcarriers (100–300 μm) has been from the perspective of crystallisation/abrasion. Particle damage is mostly related to particle-impeller and particle–particle collisions. In either case, damage increases with increasing particle size, with increasing particle concentration and with increasing agitation intensity (Nienow, 1997b). Other work has shown that freely suspended mammalian cells (size 15–20 μm) are much less sensitive to fluid dynamic stresses induced by agitation or sparged aeration than originally thought. Indeed, the impact of bursting bubble effects can be largely discounted if Pluronic F68 is included in the culture medium (Oh et al., 1992). However, those growing on microcarriers appear to be more sensitive, probably because they are attached to relatively large particles that are more prone to collisions, which might damage the cells (Nienow, 1997b).

For fibroblasts, cell damage in these circumstances has been related to the relative size of the microcarriers and the Kolmogorov microscale of turbulence (Croughan et al., 1987). Damage became significant when the microscale was about two-thirds the size of the microcarriers, or smaller (Croughan et al., 1987). In addition to these considerations, it is also important to ensure that cells are attached to the microcarriers efficiently and *a priori* it is not obvious how this should be done. Clearly, if the microcarriers in the bioreactor are not suspended during attachment, the cells will tend to attach only to the upper surface of the microcarriers. Furthermore, even if there is sufficient agitation to cause suspension and even possibly increased cell-microcarrier contacts under the action of turbulence, attachment may not occur effectively if the time of contact is too short or the turbulence too great. Recently, these phenomena were investigated by us, using hMSCs in 250 ml spinner flasks with two impeller geometries (Figure 4.3) and a well-known commercially available microcarrier, Cytodex 3 (GE Healthcare, Sweden), shown to be suitable for the expansion of both murine embryonic and

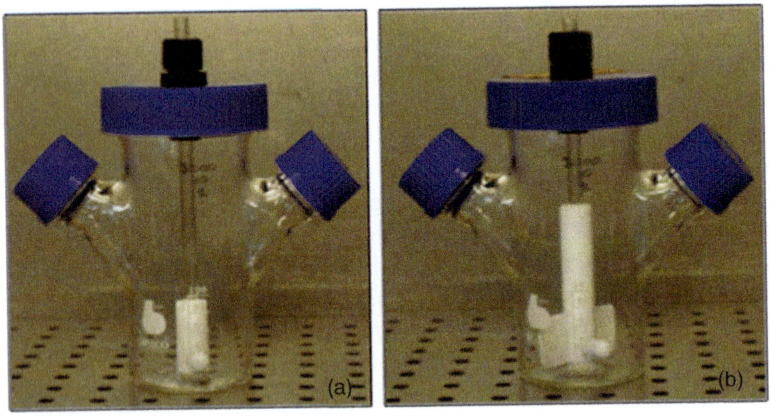

Impeller	Diameter D (mm)	Height (mm)	Swept volume $\times 10^5$ (m^3)	Impeller speed N (rps)	Power number Po	Specific energy dissipation rate ε_r $\times 10^{-4}$ ($W\ kg^{-1}$)	Estimated microscale of turbulence λ_K (μm)
G1	45	10	1.59	0.83	0.5	34	131
G2	55	30	7.13	0.5	1.0	8.8	183

Figure 4.3 The two different impeller geometries that were used in the 100 ml (working volume) spinner flasks: the first, G1 (a) consisted only of a magnetically driven horizontally suspended flea (length 45 mm; diameter 10 mm), whilst the second, G2 (b), was made up from the same suspended flea with an additional vertical Teflon paddle (30 mm high; length 55 mm). Calculation of Kolmogorov microscale of turbulence, λ_K μm. Impeller G1 (Flask A) consisted only of a magnetically driven horizontally suspended flea (length 45 mm; diameter 10 mm), whilst impeller G2 (Flask B) had an additional vertical Teflon paddle (30 mm high; length 55 mm) (Hewitt et al., 2011).

mesenchymal stem cells (Abranches et al., 2007; Frauenschuh et al., 2007).

The effects on cell expansion of impeller geometry, seeding density and microcarrier concentration were investigated (Table 4.5). For each geometry, the minimum agitation speed for microcarrier suspension, N_{JS}, was used, to ensure the microcarrier surfaces were available whilst minimizing possible agitation damage. In each case, the effect of introducing a delay in starting agitation was investigated. During such a delay, cell attachment might occur. It was found that the Cytodex 3 microcarriers were not damaged when held at the minimum speed, N_{JS}, for their suspension, using either impeller, nor was there any observable damage to the cells. The maximum cell density was achieved after 8–10 days of culture, with up to a 20-fold expansion in terms of cells per microcarrier. An increase in microcarrier concentration or seeding density generally had a deleterious or neutral effect, as previously observed for human fibroblast cultures. The choice of impeller was significant, as was incorporation of a 1-day delay before agitation to allow initial attachment of cells. The best conditions for cell expansion on the microcarriers in the flasks were 3,000 microcarriers ml^{-1} (ca. 1 g dry weight l^{-1}), a seeding density of 5 cells per microcarrier with a 1-day delay before agitation began at N_{JS} (30 rpm), using a horizontally suspended flea impeller with an added vertical paddle.

4.6 Large-scale Expansion in Conventional Stirred Tank Bioreactors

Clearly, microcarrier technology not only provides a significantly larger surface area per unit volume of bioreactor (Nienow, 2006) compared to T-flask culture, but also combines the potential ease of scalability, flexible modes of operation, process monitoring and control capability associated with bioreactor cultures that makes bioreactor culture so widespread in the biopharmaceutical industry. Microcarrier culture is commonly used as a large-scale expansion technique for the culture of adherent cells in vaccine production and was first described in 1967 where positively charged DEAE-Sephadex beads were used to culture rabbit embryonic skin cells and human embryonic lung cells (van Wezel, 1967). However, in vaccine development, the product of

Table 4.5 Summary of the mean maximum number of cells achieved per Cytodex 3 microcarrier under various conditions using both types of 100 ml spinner flasks G1 (Flask A from Figure 4.3) and G2 (i.e. Flask B from Figure 4.3) (Hewitt et al., 2011).

Microcarrier concentration (microcarrier mL^{-1})	Seeding density (cells per microcarrier)	1 day delay	Mean maximum cell number per microcarrier	Expansion factor	Mean maximum cell number $mL^{-1} \times 10^{-5}$
Impeller G1 (A)					
3000	5	×	56 ± 1	11	1.7
3000	5	√	54 ± 5	11	1.6
3000	10	×	21 ± 5	2	0.6
3000	10	√	80 ± 7	8	2.4
7500	5	×	26 ± 5	5	2.0
7500	5	√	50 ± 3	10	3.8
7500	10	×	18 ± 1	2	1.4
7500	10	√	39 ± 1	4	2.9

Impeller G2 (B)					
3000	5	×	64 ± 14	13	1.9
3000	5	√	101 ± 11	20	3.0
3000	10	×	22 ± 7	2	0.7
3000	10	√	64 ± 3	6	1.9
7500	5	×	51 ± 10	10	3.8
7500	5	√	48 ± 1	10	3.6
7500	10	×	34 ± 3	3	2.6
7500	10	√	43 ± 2	4	3.2

interest is usually an attenuated/killed virus, with little or no attention paid to the cell during downstream processing. Therefore a number of challenges in adapting this technology for use in the large-scale expansion of anchorage-dependent stem cells such as hMSCs remain.

With over 380 clinical trials taking place involving the use of hMSCs (Heathman et al., 2015), there is now a clear need for the development of platform technologies for their large-scale culture, which would yield them in sufficient quantity and quality. Studies have taken place investigating the culture of hMSCs on microcarriers with a range of commercially available microcarriers being used, including Cytodex-1 (Frauenschuh et al., 2007; Schop et al., 2008), Cytodex-3 (Hewitt et al., 2011), Cultispher-S (Eibes et al., 2010), Cultispher-G (Sun et al., 2010) and SoloHill Plastic P102-L (Dos Santos et al., 2011). However, in all of these studies, the largest volume used was no more than 200 ml and it is unclear as to whether samples of microcarrier suspension greater than 2 ml were ever harvested. Detachment of cells from the microcarrier surface and subsequent retention of cell quality is equally as important as cell attachment and proliferation, given that the product of interest for cell-based therapies is the cells. Therefore it is of paramount importance to consider cell-harvesting strategies from the outset, so as to ensure a viable, holistic bioprocess. In our work (Rafiq et al., 2013a), fully functional hMSCs isolated from bone marrow were cultured for the first time in fully controllable 5 L bioreactors with a working volume of 2.5 L for 12 days on Plastic microcarriers (Figure 4.4). These microcarriers were selected based on their suitability for the culture of hMSCs relative to a number of other microcarriers as identified by us previously (Rafiq et al., 2016), as well as by other groups (Dos Santos et al., 2011), in addition to their amenability for good manufacturing practice (GMP) bioprocesses, given their xeno-free composition.

Whilst it would be possible to control the pH and dO_2 levels in the bioreactors, it was decided that this investigation would be used to monitor and identify the change in pH and dO_2 over the course of the culture, as there is little published work that identifies the optimal pH and dO_2 levels for microcarrier hMSC culture. Online monitoring of pH and dO_2 is a rarity in hMSC culture (monolayer or microcarrier). Thus being able to shift from an unmonitored to an online monitored process with the possibility of control is a significant step forward. Cells were also cultured in duplicate 100 ml spinner flasks in parallel

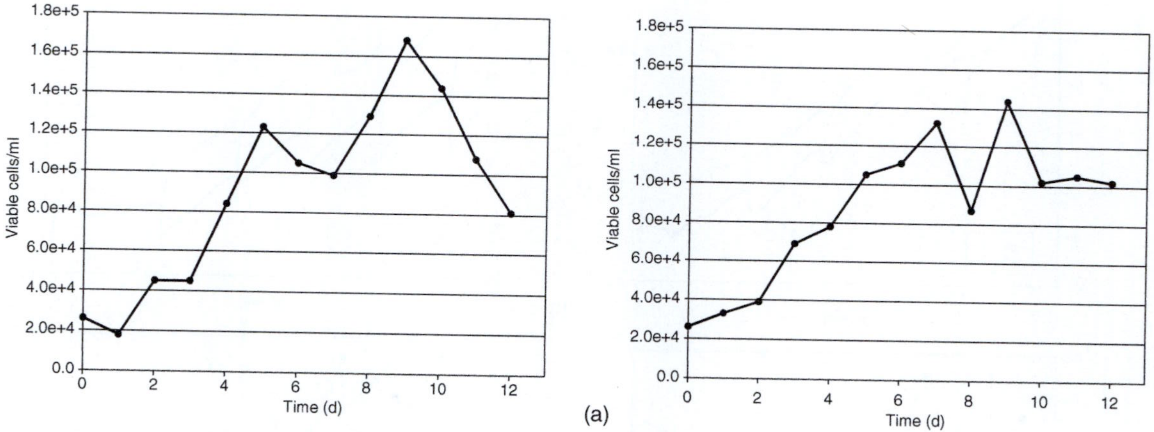

Figure 4.4 The growth of hMSCs in 5 L bioreactor vessels over a period of 12 days. (a) Viable cells/ml readings for bioreactor 1 (left) and bioreactor 2 (right); (b) nutrient and metabolite concentrations for bioreactor 1 (left) and bioreactor 2 (right); (c) pH and dO_2 readings for bioreactor 1 (left) and pH and dO_2 readings for bioreactor 2 (right) (Rafiq et al., 2013a).

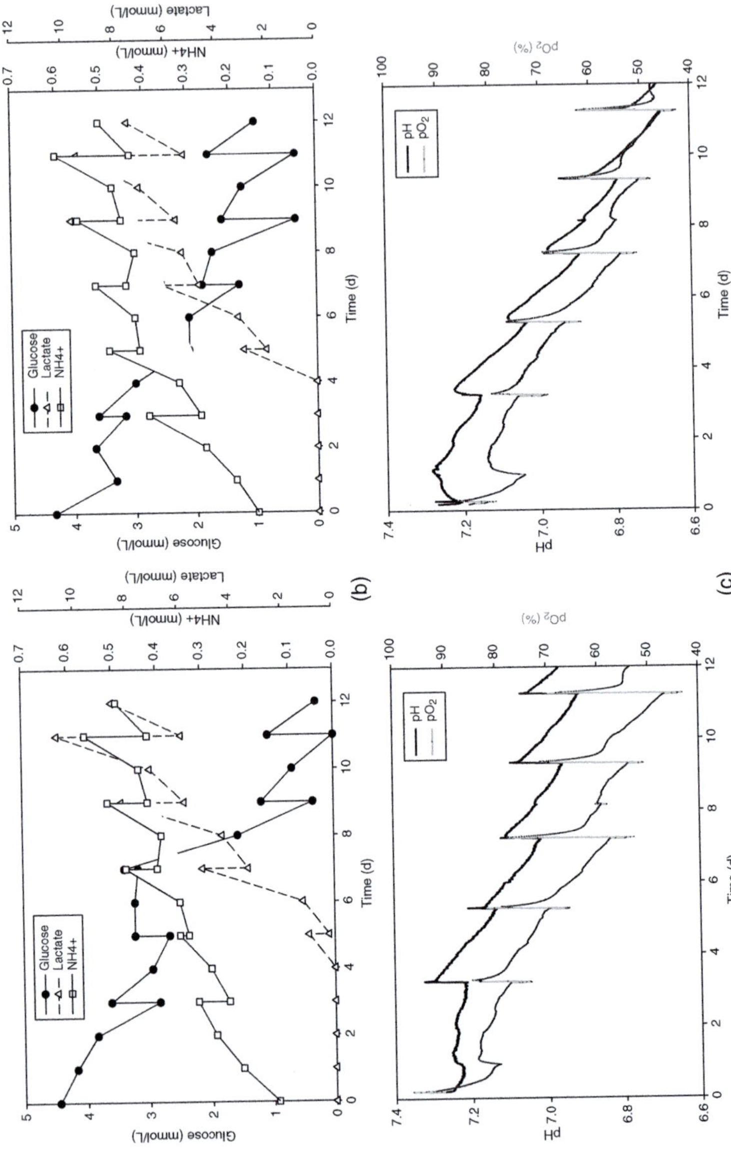

Figure 4.4 (Continued)

with the bioreactors for comparison. Cells were then harvested *via* a potentially scalable detachment protocol that allowed for the successful detachment of hMSCs from the cell-microcarrier suspension (Nienow et al., 2014).

Over 12 days, the dO_2 concentration was more than 45% of saturation and the pH between 7.2 and 6.7 giving a maximum cell density in the 5 L bioreactor of 1.79×10^5 cells/ml; this represent a 6-fold expansion of the hMSCs, equivalent to that achievable from 65 fully-confluent T-175 flasks. Importantly this would equate to that produced by one CompacT Select automated platform (TAP Biosytems, UK) if the full 5 L volume had been used. During the culture period, the average specific O_2 uptake of the cells in the 5 L bioreactor was 8.1 fmol/cell/h and, in all cases, the 5 L bioreactors outperformed the equivalent 100 ml spinner-flasks run in parallel with respect to cell yields and growth rates. In addition, yield coefficients, specific growth rates and doubling times were calculated for all systems. Neither the upstream nor downstream bioprocessing unit operations had a discernible effect on cell quality, with the harvested cells retaining their immunophenotypic markers, key morphological features and differentiation capacity. However, it is recognised that the productivity of even the best bioreactor run, equivalent to that produced manually from 65 confluent T-175 flasks, was insufficient to meet the demand for the high numbers of cells required for a single cell-based therapy (Rafiq et al., 2013a).

Nevertheless the results are promising, given the opportunity for further optimisation of the culture conditions including feed strategy (e.g. perfusion) and the harvesting step as well as the advantage of scalability and online control with a fully instrumented bioreactor system. Importantly, the improved performance in the 5 L baffled bioreactors compared to the spinner flasks is particularly encouraging, because such configurations are the easiest to scale-up (Nienow, 2006), and there are physical studies of reactors and bioreactors in the literature to suggest that further increases in scale with such geometries are likely to be successful.

4.7 Cell Recovery from Microcarriers

As stated previously, the choice of microcarrier depends on several factors, including cell attachment and proliferation, and whether they are xeno-free. However, in the majority of cultivation studies listed above, there has been little focus on the harvesting procedure, or consideration of the ability to effectively harvest should the process increase in scale (Chen et al., 2013). Yet, detachment or dissociation of cells from the microcarrier surface and subsequent retention of cell quality is equally as important as cell attachment and proliferation, given that the product of interest for cell therapies is the cell itself. It is also important to recognise that, independent of whether the cells are to be reused for further expansion following a passage or stored prior to use for therapeutic or other purposes, harvesting involves two steps. Firstly, the cells are detached from microcarriers to produce a cell–microcarrier suspension; and secondly, a further separation step leaving the cells in suspension without the microcarriers present. Therefore we have employed the term "harvesting" to denote this two-step process of recovering the hMSCs free from microcarriers.

The term "detachment" or "dissociation" is used just for the first step of removing the cells from the microcarrier surface, and the second step is designated "separation" and involves the use of techniques such as filtration or centrifugation. These issues will only be exacerbated as the expansion scale increases and therefore it is crucial to consider cell-harvesting strategies from the outset so as to ensure a viable, complete bioprocess. Typically, cell detachment from microcarriers is by enzymatic digestion (Dos Santos et al., 2013). Weber et al. (2007) studied detachment using different enzymes and a range of microcarriers (Biosilon, Cytodex1, Cytodex 3, RapidCell and P102-L). They expanded hMSC-TERT cells in spinner flasks and then detached the cells by exposing them to trypsin, accutase or collagenase or a trypsin-accutase mixture generally for 6 min. The cells cultivated on Biosilon, RapidCell and P102-L were detached successfully, especially with the trypsin, and trypsin-accutase mixture achieving a high cell yield and viabilities near 100%. The results using collagenase were generally poor for Cytodex 1 and Cytodex 3 with any of the enzyme solutions, where even after exposure for up to 10 min, significant detachment of the cells was not obtained. After another spinner flask expansion and harvest, the cells were successfully differentiated to adipocytes. Though a very interesting paper, the scale at which the harvesting was undertaken was only 5 mL, very small compared to current and predicted future needs. Others have suggested that a non-enzymatic harvesting procedure would be preferable, so as to avoid potential cell damage or change in the immunophenotype of the cell (Jang et al., 2012; Potapova et al., 2008).

Work by Yang et al. (2010) has highlighted the possibility of harvesting hMSCs without the need for enzymatic digestion by using a modified Cytodex-3 microcarrier coated with a thermo responsive polymer. The hMSCs were expanded on these microcarriers, after which the culture temperature was lowered to 32°C, where it was suggested that hMSCs were able to detach from the microcarriers. However, this method involves a significant temperature drop, difficult to achieve quickly at the large scale and the addition of an extra process step (coating of the microcarrier with the polymer), which in effect makes the innate surface properties of a microcarrier redundant, which may prove problematic if a process has been designed to take advantage of its original surface properties. It may also prove difficult to obtain an even coating of the thermo responsive polymer across the entire surface of the microcarrier, especially on scale-up. Enzyme-free dissociation buffers also exist but are rarely reported to be used for routine passaging of cells and some formulations may not be suitable for use with hMSCs. Indeed, one study using a Gibco-brand enzyme-free dissociation buffer to detach hMSCs grown on tissue culture plastic found that cell viability following detachment was significantly reduced compared to trypsin controls (~70% compared to >90%, respectively) and that cells also failed to re-attach to tissue culture plastic (Heng et al., 2009).

It was decided to develop a larger-scale protocol based on using agitation to speed up the process. This approach is not an obvious one, as generally there is a concern that excessive agitation damages cells, particularly animal cells because of the lack of a cell wall (Nienow 2006). However, it is also recognised that excessive exposure to the types of enzymes used in detachment can cause processing problems, even when the cells are not themselves the product (Bluml, 2007). The amount of damage depends on the type and concentration of the enzyme and the time of exposure (Jang et al., 2012). On the other hand, there is increasing evidence that animal cells are less sensitive to mechanical stress than originally thought, both with respect to cell viability and density and protein quality (Nienow et al., 2013).

It was therefore decided to use trypsin-EDTA for detachment, but to enhance it by increased mechanical stress from agitation for a shorter time than that suggested in the manufacturer's protocol, to minimise damage to the cells whether from stress or enzymatic action. This new detachment method is based on theoretical concepts originally developed for secondary nucleation due to agitation. Based on this theory,

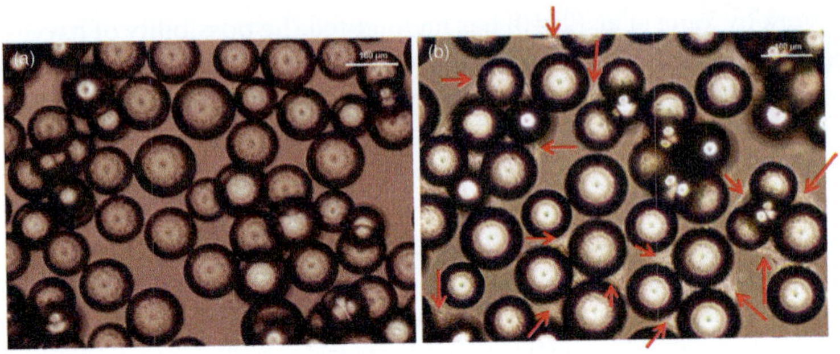

Figure 4.5 Phase-contrast image of microcarrier and cell suspension after dissociation during the novel harvesting procedure. Magnification 10x. Arrows: (a) Single hMSC successfully detached from microcarrier; and (b) hMSCs still attached to microcarrier (Nienow et al., 2014).

it is suggested that a short period (here 7 min) of intense agitation in the presence of a suitable enzyme should detach the cells from the relatively large microcarriers. In addition, once detached, the cells should not be damaged because they are smaller than the Kolmogorov microscale. Detachment was then successfully achieved for hMSCs from two different donors grown on SoloHill microcarriers in a 5 L bioreactor (Nienow et al., 2014). In both cases, harvesting was completed by separating cells from microcarriers using a Steriflip® vacuum filter (Figure 4.5). The overall harvesting efficiency was more than 95% and after harvesting, the cells maintained all the attributes expected of hMSCs. The underlying theoretical concepts suggest that the method is scalable and this aspect is discussed elsewhere too.

4.8 Conclusions

It is clear that for autologous therapies a scale-out approach may be appropriate using automated cell culture platforms as demonstrated successfully by others, but for allogeneic therapies it is likely that larger-scale bioprocesses will be required. Our work has demonstrated the feasibility of a scalable microcarrier process for the expansion and harvest of hMSCs in conventional stirred tank bioreactors. The integration of multiple process unit operations is an important step in developing a microcarrier-based expansion process capable of meeting the lot sizes required for clinical applications. hMSC identity and quality have been maintained throughout every unit operation of this integrated process, culminating with the successful harvest of hMSCs from the microcarriers and spent culture broth at the end of the expansion process. In addition, we have quantified the important bioprocessing metabolites under static and agitated culture conditions that can be used to inform further scale-up. However, it is our contention that we are still in the "foothills" for stem cell bioprocess development. Globally there are relatively very few academic or industrial groups working on developing such bioprocesses, when compared to the numbers of groups working on the basic biology of such cells. If the therapeutic potential of stem cells is ever to be realised, then we need fully defined scalable upstream expansion processes, downstream separation technologies and suitable cell preservation strategies that will de-couple cell expansion from delivery to the patient. In order to do this, there needs to be greater collaboration between engineers, scientists and medical professionals.

References

Abranches, E., Bekman, E., Henrique, D. and Cabral, J.M. (2007) Expansion of mouse embryonic stem cells on microcarriers. *Biotechnol. Bioeng.*, **96**, 1211–21.

Bluml, G. (2007) Microcarrier cell culture technology, in *Animal Cell Biotechnology* (ed. R. Pörtner), Humana Press, New York, US.

Boswell, C.D., Nienow, A.W., Gill, N.K. *et al.* (2003) The impact of fluid mechanical stress on *Saccharomyces cerevisiae* cells during continuous cultivation in an agitated, aerated bioreactor: Its implication for mixing in the brewing process and aerobic fermentations. *Food and Bioproducts Processing*, **81**, 23–32.

Caplan, A.I. (2007) Adult mesenchymal stem cells for tissue engineering versus regenerative medicine. *J Cell Physiol*, **213**, 341–7.

Chen, A.K.-L., Reuveny, S. and Oh, S.K.W. (2013) Application of human mesenchymal and pluripotent stem cell microcarrier cultures in cellular therapy: Achievements and future direction. *Biotechnology Advances*, **31**, 1032–46.

Cherry, R.S. and Papoutsakis, E.T. (1986) Hydrodynamic effects on cells in agitated tissue culture reactors. *Bioprocess Engineering*, **1**, 29–41.

Croughan, M.S., Hamel, J.F. and Wang, D.I. (1987) Hydrodynamic effects on animal cells grown in microcarrier cultures. *Biotechnol Bioeng.*, **29**, 130–41.

Croughan, M.S., Hamel, J.F. and Wang, D.I. (1988) Effects of microcarrier concentration in animal cell culture. *Biotechnol Bioeng.*, **32**, 975–82.

Croughan, M.S., Sayre, E.S. and Wang, D.I. (1989) Viscous reduction of turbulent damage in animal cell culture. *Biotechnol Bioeng.*, **33**, 862–72.

Dos Santos, F., Andrade, P.Z., Eibes, G. *et al.* (2011) *Ex vivo* expansion of human mesenchymal stem cells on microcarriers. *Methods Mol. Biol.*, **698**, 189–98.

Dos Santos, F.F., Andrade, P.Z., Da Silva, C.L. and Cabral, J.M. (2013) Bioreactor design for clinical-grade expansion of stem cells. *Biotechnol. J.*, **8**, 644–54.

Eibes, G., Dos Santos, F., Andrade, P.Z. *et al.* (2010) Maximizing the *ex vivo* expansion of human mesenchymal stem cells using a microcarrier-based stirred culture system. *J. Biotechnol.*, **146**, 194–7.

Enfors, S.O., Jahic, M., Rozkov, A. *et al.* (2001) Physiological responses to mixing in large-scale bioreactors. *J. Biotechnol.*, **85**, 175–85.

Estrada, J.C., Albo, C., Benguría, A. *et al.* (2011) Culture of human mesenchymal stem cells at low oxygen tension improves growth and genetic stability by activating glycolysis. *Cell Death and Differentiation*, **19**, 743–55.

Frauenschuh, S., Reichmann, E., Ibold, Y. *et al.* (2007) A microcarrier-based cultivation system for expansion of primary mesenchymal stem cells. *Biotechnol. Prog.*, **23**, 187–93.

Grayson, W.L., Zhao, F., Bunnell, B. and Ma, T. (2007) Hypoxia enhances proliferation and tissue formation of human mesenchymal stem cells. *Biochem Biophys. Res. Commun.*, **358**, 948–53.

Heathman, T.R.J., Nienow, A.W., Mccall, M.J. *et al.* (2015) The translation of cell-based therapies: clinical landscape and manufacturing challenges. *Regenerative Medicine*, **10**, 49–64.

Heng, B.C., Cowan, C.M. and Basu, S. (2009) Comparison of enzymatic and non-enzymatic means of dissociating adherent monolayers of mesenchymal stem cells. *Biol. Proced. Online*, **11**, 161–9.

Hewitt, C.J. (2003) Special topic issue – Brewing: an ancient process benefiting from modern scientific and engineering techniques. *Food and Bioproducts Processing*, **81**, 1.

Hewitt, C.J. and Nienow, A.W. (2007) The scale-up of microbial batch and fed-batch fermentation processes. *Advances in Applied Microbiology*, **62**, 105–35.

Hewitt, C.J. and Solomons, G.L. (1996) The production of α-amylase (E.C.3.2.1.1.) by *Bacillus amyloliquefaciens*, in a complex and a totally defined synthetic culture medium. *Journal of Industrial Microbiology*, **17**, 96–9.

Hewitt, C.J., Boon, L.A., Mcfarlane, C.M. and Nienow, A.W. (1998) The use of flow cytometry to study the impact of fluid mechanical stress on *Escherichia coli* W3110 during continuous cultivation in an agitated bioreactor. *Biotechnol. Bioeng.*, **59**, 612–20.

Hewitt, C.J., Isailovic, B., Mukenwa, N.T. and Nienow, A.W. (2006) Insect and animal cell technology, in *Basic Biotechnology*, 3rd edn (eds C. Ratledge and B. Kristiansen), Cambridge University Press, Cambridge, UK.

Hewitt, C.J., Lee, K., Nienow, A.W. *et al.* (2011) Expansion of human mesenchymal stem cells on microcarriers. *Biotechnol Lett.*, **33**, 2325–35.

Hung, S.P., Ho, J.H., Shih, Y.V. *et al.* (2012) Hypoxia promotes proliferation and osteogenic differentiation potentials of human mesenchymal stem cells. *Journal of Orthopaedic Research*, **30**, 260–6.

Ibrahim, S. and Nienow, A.W. (2004) Suspension of microcarriers for cell culture with axial flow impellers. *Chemical Engineering Research and Design*, **82**, 1082–8.

Jones, S., Mckee, S.D. and Levine, H.L. (2012) Emerging challenges in cell therapy manufacturing. *Bioprocess International*, **10**, 4–7.

Jang, S., Panchalingam, K.M., Wuerth, R.D. *et al.* (2012) Large-scale production of human mesenchymal stem cells for clinical applications. *Biotechnology and Applied Biochemistry*, **59**, 106–20.

Kumar, N., Gammell, P., Meleady, P. *et al.* (2008) Differential protein expression following low temperature culture of suspension CHO-K1 cells. *BMC Biotechno.l*, **8**, 42.

Lax, E. (2005) *The Mould In Dr Florey's Coat: The Remarkable True Story of the Penicillin Miracle*, London, UK, Abacus.

Mason, C. and Hoare, M. (2006) Regenerative medicine bioprocessing: The need to learn from the experience of other fields. *Regenerative Medicine*, **1**, 615–23.

Nienow, A.W. (1997a) The mixer as a reactor – liquid/solid systems, in *Mixing in the Process Industries*, 2nd edn (eds N. Harnby, M.F. Edwards and A.W. Nienow), Butterworth Heinemann, London.

Nienow, A.W. (1997b) The suspension of solid particles, in *Mixing in the Process Industries*, 2nd edn (eds N. Harnby, M.F. Edwards and A.W. Nienow), Butterworth Heinemann, London.

Nienow, A.W. (2006) Reactor engineering in large-scale animal cell culture. *Cytotechnology*, **50**, 9–33.

Nienow, A.W. (2014) Letter to the Editor on the Paper by Sieck, J.B. et al., "Development of a Scale-down model of hydrodynamic stress to study the performance of an industrial CHO Cell line under simulated production scale bioreactor conditions." *J. Biotechnol.*, 164(2013), 41–49. *J. Biotechnol.*, **171**, 82–4.

Nienow, A.W., Rielly, C.D., Brosnan, K. *et al.* (2013) The physical characterisation of a microscale parallel bioreactor platform with an industrial CHO cell line expressing an IgG4. *Biochem. Eng. J.*, **76**, 25–36.

Nienow, A.W., Rafiq, Q.A., Coopman, K. and Hewitt, C.J. (2014) A potentially scalable method for the harvesting of hMSCs from microcarriers. *Biochemical Engineering Journal*, **85**, 79–88.

Oh, S.K.W., Nienow, A.W., Al-Rubeai, M. and Emery, A.N. (1992) Further studies of the culture of mouse hybridomas in an agitated bioreactor with and without continuous sparging. *J. Biotechnol.*, **22**, 245–70.

Oh, S.K.W., Nienow, A.W., Al-Rubeai, M. and Emery, A.N. (1989) The effects of agitation intensity with and without continuous sparging on the growth and antibody production of hybridoma cells. *Journal of Biotechnology*, **12**, 45–61.

Oh, S.K.W., Chen, A.K., Mok, Y. *et al.* (2009) Long-term microcarrier suspension cultures of human embryonic stem cells. *Stem Cell Res*, **2**, 219–30.

Potapova, I.A., Brink, P.R., Cohen, I.S. and Doronin, S.V. (2008) Culturing of human mesenchymal stem cells as three-dimensional aggregates induces functional expression of CXCR4 that regulates adhesion to endothelial cells. *Journal of Biological Chemistry*, **283**, 13100–7.

Rafiq, Q.A. and Hewitt, C.J. (2015) Cell therapies: Why scale matters. *Pharmaceutical Bioprocessing*, **3** (2), 97–9.

Rafiq, Q.A., Brosnan, K.M., Coopman, K. *et al.* (2013a) Culture of human mesenchymal stem cells on microcarriers in a 5 l stirred-tank bioreactor. *Biotechnol. Lett.*, **35**, 1233–45.

Rafiq, Q.A., Coopman, K. and Hewitt, C.J. (2013b) Scale-up of human mesenchymal stem cell culture: Current technologies and future challenges. *Current Opinion in Chemical Engineering*, **2**, 8–16.

Rafiq, Q.A., Coopman, K., Nienow, A.W. and Hewitt, C.J. (2013c) A quantitative approach for understanding small-scale human mesenchymal stem cell culture - Implications for large-scale bioprocess development. *Biotechnol. J.*, **8**, 459–71.

Rafiq, Q.A., Coopman, K., Nienow, A.W. and Hewitt, C.J. (2016) Systematic microcarrier screening and agitated culture conditions improves human mesenchymal stem cell yield in bioreactors. *Biotechnology Journal*, **11**, 473–86.

Schop, D., Janssen, F.W., Borgart, E. *et al.* (2008) Expansion of mesenchymal stem cells using a microcarrier-based cultivation system: growth and metabolism. *J. Tissue Eng. Regen. Med*, **2**, 126–35.

Sun, L.Y., Hsieh, D.K., Syu, W.S. *et al.* (2010) Cell proliferation of human bone marrow mesenchymal stem cells on biodegradable microcarriers enhances *in vitro* differentiation potential. *Cell Prolif.*, **43**, 445–56.

Thomas, C.R. (1990) Problems of shear in biotechnology, in *Chemical Engineering Problems in Biotechnology* (ed. M.A. Winkler), Elsevier Applied Sciences, UK.

Thomas, R.J., Chandra, A., Liu, Y. *et al.* (2007) Manufacture of a human mesenchymal stem cell population using an automated cell culture platform. *Cytotechnology*, **55**, 31–9.

Thomas, R.J., Anderson, D., Chandra, A. *et al.* (2009) Automated, scalable culture of human embryonic stem cells in feeder-free conditions. *Biotechnology and Bioengineering*, **102**, 1636–44.

Van Wezel, A.L. (1967) Growth of cell-strains and primary cells on micro-carriers in homogeneous culture. *Nature*, **216**, 64–5.

Want, A.J., Nienow, A.W., Hewitt, C.J. and Coopman, K. (2012) Large-scale expansion and exploitation of pluripotent stem cells for regenerative medicine purposes: beyond the T-flask. *Regen. Med.*, **7**, 71–84.

Weber, C., Pohl, S., Pörtner, R. *et al.* (2007) Expansion and harvesting of hMSC-TERT. *The Open Biomedical Engineering Journal*, **1**, 38.

Wurm, F. (2013) CHO Quasispecies – Implications for manufacturing processes. *Processes*, **1**, 296–311.

Yang, H.S., Jeon, O., Bhang, S.H. *et al.* (2010) Suspension culture of mammalian cells using thermosensitive microcarrier that allows cell detachment without proteolytic enzyme treatment. *Cell Transplantation*, **19**, 1123–32.

5

Challenges of Scale-up of Cell Separation and Purification Techniques

Marieke A. Hoeve[1,2], *Paul A. De Sousa*[1,2] *and Nicholas A. Willoughby*[3]

[1] *Centre for Clinical Brain Sciences, University of Edinburgh, Edinburgh, UK*
[2] *MRC Centre for Regenerative Medicine, University of Edinburgh, Edinburgh, UK*
[3] *Institute of Biological Chemistry, Biophysics and Bioengineering, School of Engineering and Physical Sciences, Heriot-Watt University, Edinburgh, UK*

5.1 Introduction

5.1.1 Cell Separation for Cell-based Therapeutics

Progress in stem cell research and regenerative medical treatments is leading towards the development of more complex cell-based therapeutic products which promise to impact on many aspects of healthcare. Cell-based products create new issues for bio-production and processing, requiring a re-think of current manufacturing strategy. A key consideration in the manufacturing of any product is its purity and freedom from contamination. In the case of stem cell derived therapeutics, this encompasses freedom from unknown or adventitious pathogens, through to the homogeneity of cell populations. Control of cellular homogeneity may be important throughout a manufacturing process, to ensure robustness and reproducibility of protocols and purity of the intended end product, regardless of whether it is used *in vitro* for discovery or screening applications or for transplantation *in vivo* to treat degenerative disease or damaged tissues. A critical concern for cell therapies, especially cells derived from embryo stem cells or cells induced into an embryo stem cell-like state, will be the absence of undifferentiated cell populations in the final cellular

Bioprocessing for Cell Based Therapies, First Edition. Edited by Che J. Connon.

product intended for transplantation. These cells have the potential to form teratomas, benign tumours capable of interfering with tissue physiology, or worse, teratocarcinomas, with metastatic potential. Consequently, there exists a pressing manufacturing need for effective separation technologies to produce a safe end product.

5.1.2 Separation Methodology Design

Design of separation technology suitable for cell manufacture needs to be instructed from the point of view of the intended product (i.e. composition, purity and viability), as well as the process as a whole, including sample preparation and handling, manufacturing, running costs, ease of operation (skills required) and Good Manufacturing Practice compatibility.

When developing a particular cell separation method, various aspects need to be taken into account. Firstly, the source of starting materials and intended use of the end product, which will dictate the required sample purity and consequently selection efficiency of the method. In this context, new challenges have arisen due to recent achievements in the stem cell field, in particular the need for technologies to deplete tumorigenic cells after differentiation of pluripotent stem cells for therapeutic use. Closely related to these considerations is the positive or negative nature of the cell selection: positive selection, entailing harvesting of wanted cells, whilst with negative selection unwanted cells are harvested for discarding.

The choice of methodology should furthermore be instructed by the point in the cell manufacturing process at which the purification is to be applied, as this will dictate the suitability of a method, and impact on methods used to culture cells, media formulation and the required scale of cell production.

5.1.3 Objective of this Chapter

As this chapter is part of a book on cell-based therapies, we will focus on methodologies suitable for mammalian cell separation as opposed to purification of other biological entities like particles, proteins and non-mammalian cells. We will describe reported separation methods that may lend themselves to large-scale cell separation (through scale-up rather than scale-out). Easily scalable separation techniques will doubtless become increasingly important as regenerative medicine develops, and particularly as allogeneic therapies become more common. Currently applied separation techniques, whilst effective in terms of cell isolation, are severely limited due to the need to label target cells to effect isolation and by the potential maximal scale of operation. Yet, scalability of cell separation platforms is crucial for bringing the promise of cell-based therapies to the patient, by improving the following:

5.1.3.1 Cell Yield

The number of cells required to treat a disease will be context dependent, ranging from quantities readily produced manually using conventional culture vessels (e.g. production of $\sim 1.5 \times 10^8$ retinal pigmented cells per single 6-well tissue culture plate, sufficient to treat >50 patients with macular degeneration by injection of $\sim 10^5$ retinal pigmented epithelial cells into an eye (Schwartz et al., 2012). to vastly greater quantities requiring automated culture systems which have yet to be engineered (e.g. $\sim 2 \times 10^{12}$ red blood cells per unit of blood for transfusion). Cell yields will also be determined by desired batch production volumes to accommodate requirements for clinical testing, repeat dosing and favourable health economics.

5.1.3.2 Standardisation

Regulatory authorisation of cell usage in clinical trials and subsequent patient treatments necessitate standardised and validated production protocols. Cell products may be produced fresh for immediate use or as cryostored “off the shelf” products.

5.1.3.3 Economical viability

The health economics of cell production and the therapeutic efficacy of the therapy on which it is based must be competitive with existing non-cellular treatments or management of a disease and its symptoms. The potential for large-scale manufacture will enable developers to more easily recover the costs of product development, optimisation and testing, and to ultimately produce affordable, commercially viable therapeutics.

Here we will describe the working principles of a number of cell separation methods that have been reported in the literature or are currently applied for mammalian cell separation. Each method is set against parameters that need to be considered when developing large-scale cell purification for the clinic.

5.2 Scalable Cell Separation for Cell Therapy

Scale-up of cell manufacture can be achieved horizontally, through scale-out (serial production with multiple devices/production sites) and vertically, through scale-up (increasing the volumetric or cell number throughput of a single device). Cell therapeutics that require or would benefit from scalability through scale-up are off-the-shelf therapies based on the use of pluripotent stem cells (embryonic, mesenchymal or induced pluripotent stem cells) or differentiated cell populations, while scale-out manufacture might suffice for patient-specific therapies. We propose at least six parameters should be considered when assessing the suitability of cell separation techniques for large-scale cell manufacture by means of scale-up production.

5.2.1 Label Requiring versus Label-free Separation

The two main cell separation methods currently widely used for research and clinical (mainly diagnostics) practices are flow cytometric fluorescent-activated cell sorting (FACS) (Figure 5.1) and magnetic cell separation technologies (Figure 5.2). The latter include Miltenyi's magnetic-activated cell sorting (MACS®), Invitrogen's Dynabeads®, BDs IMag® and StemCell™Technology's EasySep® systems. Fluorescent- and magnetic-activated cell sorting methods are both based on the recognition and labelling of cell-associated epitopes located inside or on the surface of cells. Detection normally relies on targeted epitopes using high-affinity probes including antibodies, peptides or nucleic acids directly or secondarily labelled with photo-excitable and light emitting molecules or nano-or micro-sized particles. Though proven very useful for a variety of applications and having the potential to yield highly pure cell populations (>95%), the major drawback of cell labelling strategies is the potential for recognition of the targeted epitope or the labelling itself to compromise cellular viability or functions. This is especially so for labelling of intracellular epitopes requiring internalisation of probes. Cell labelling also adds process time and costs, and carries a greater risk of inadvertent contamination of cell products introduced via labelling reagents or cell processing, such as during flow cytometric separation in communally used equipment. In addition, the success of high affinity labelling is dependent on the strength and exclusivity of expression of the targeted epitope used to discriminate cell populations.

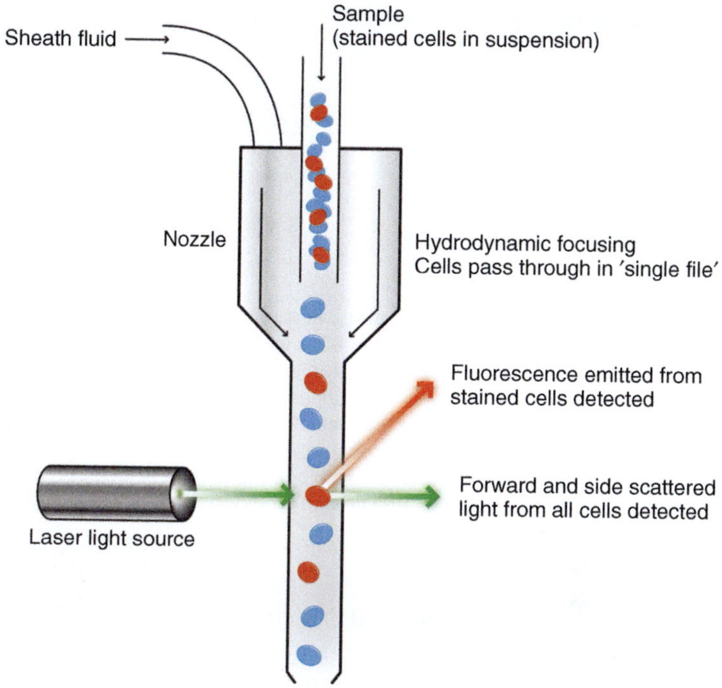

Figure 5.1 Principle of fluorescence-activated cell sorting (FACS). After a single cell suspension is incubated with one or more fluorescently labelled antibodies, a sample is diluted in suspension buffer and run through the cytometer. Cells are hydrodynamically focused by sheath fluid added through a very small nozzle. The tiny "stream" of fluid takes the cells past laser beam(s) one cell at a time. A number of detectors detect the light scattered from the cells/particles as they go through the beam: one in front of the light beam (Forward Scatter or FS) and several to the side (Side Scatter or SS). Fluorescent detectors are used for the detection of fluorescence emitted from positively stained cells/particles (image provided courtesy of Abcam Inc, image copyright © 2016 Abcam).

Moreover, in the case of positive selection, the target cell will carry labels after the procedure. Subsequent removal of these labels could trigger signalling pathways, thus affecting the cell's physiological state, while incomplete removal could interfere with downstream applications such as injection into humans. But even in the case of negative selection, label-requiring methods are not optimal for many clinical applications, as antibody and label (dyes, magnetic particles) have been introduced into the sample at the start of the procedure, and may

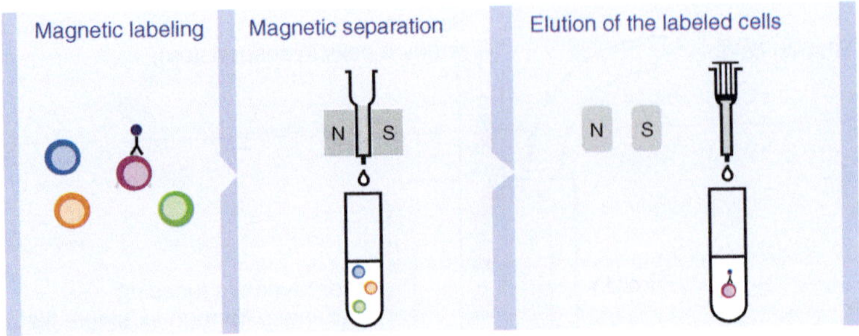

Figure 5.2 Principle of magnetic-activated cell sorting (MACS). Cells in a single-cell suspension are magnetically labelled with microbeads and applied onto a column placed in a magnetic separator. Unlabelled cells pass through, while the magnetically labelled cells are retained within the column. The flow-through can be collected as the unlabelled cell fraction. After a short washing step, the column is removed from the separator, and the magnetically labelled cells are eluted from the column with a plunger. Thus, both labelled and unlabelled cell fractions can be isolated (image provided courtesy of Miltenyi Biotec, image copyright © 2014 Miltenyi).

still be present in the enriched and depleted cell fractions. Thus, cell preparations obtained through label-requiring methods may necessitate assessment of toxicity limits prior to use in humans, rendering the cells unsuitable for clinical use until proven safe.

Although new label-requiring cell separation methods are currently being developed, the demand for cell selection techniques compatible with clinical application has resulted in the development of a large variety of label-free cell separation technologies. Whilst comparatively free of the above-described drawbacks, label-free cell separation, utilising cell biochemical or biomechanical properties such as size, polarity, charge, elasticity or deformability, may be too general, thereby compromising cell purity, selectivity and yield. Currently centrifugation and filtration are the only label-free cell purification techniques used regularly in a commercial or clinical environment. Both are reasonably effective but constitute crude techniques, primarily limited in their ability to separate cells with significantly differing physical properties (size and density). Both of these label-free methods can be scaled up, although centrifugation at larger scale becomes problematic, as larger continuous centrifuges subject cells to relatively high solid–liquid shear forces. The use of either method is most common in the separation and fractionation of blood cells and products, although density gradient centrifugation is also common practice in islet preparation for transplantation.

Label-free cell separation techniques currently in research use or under development are mostly derived from the field of biomicrofluidics, and have thus far primarily been designed for small-scale use, such as cancer diagnostics and drug monitoring rather than large-scale cell separation for therapeutics. To offer significant progress in this field, new techniques should aspire to be easily scalable, whilst offering improved selectivity compared to centrifugation and filtration, especially where target and contaminant properties only vary slightly.

5.2.2 Active versus Passive Method

Active cell separation techniques require a force for functionality, while passive techniques rely solely on the geometric design of the device to achieve cell separation. This distinction is important as it can impact on the type and amount of shear stress exerted on the cells during the separation process. This shear stress needs to be taken into consideration when deciding the suitability of a method in light of downstream applications.

5.2.3 Isolated Purification (Including Off-the-Shelf) versus Embedded Integrated Process

Production of cells for clinical applications can be divided into two categories, depending on the intended use of the end product: "patient-specific" and "off-the-shelf". Patient specific therapeutics, for example autologous cell therapies or HLA-matching strategies, can be produced in isolation in relatively small quantities, whereas cells for "off-the-shelf" therapies should be produced in large quantities at one or more specialised production facilities for subsequent distribution to clinical facilities for administration.

5.2.4 Low versus High Resolution

Resolution requirements for the separation platform will be largely dictated by the nature of the starting population. Low resolution might suffice if the starting population consists of distinctly different cells (e.g. as would be the case for red blood cell or plasma products), whilst populations made up of very similar cells or containing rare cells would demand high resolution (e.g. selection or removal of hematopoietic stem cells and cancer cells).

5.2.5 Open versus Closed Systems

Open versus closed systems for cell processing refers to the extent to which cells are exposed to the ambient environment, with a "closed" environment limited by the boundaries of the vessel or device the cells are contained in. While these boundaries may be gas permeable, they otherwise constitute barriers to pathogens or other non-gas permeable contaminants in the atmosphere. To meet regulatory standards for safety, cellular processing in open systems must be performed in environments (i.e. laminar flow hoods, "clean rooms"), which control and limit airborne contaminants below that normally detected in ambient atmosphere. The latter is assessed by monitoring particle counts in air during cell processing, and by assessing microbiological contaminants after processing by means of colony growth on soft agar plates exposed to the environment during processing. Creating, operating and maintaining such environments is very costly. Environmental specifications for operators of closed systems are substantially lower, thereby reducing costs, although this can be offset by the cost of maintaining a closed system if it is reusable. It is broadly perceived that the health economics of a cell therapy can be improved by usage of closed systems for cell bioprocessing, especially if their usage can be automated. The extent to which this is true is determined by both the scale of activity and the set-up costs associated with handling cells in closed systems. In principle, however, cell separation in closed systems is more likely to offer advantages in product consistency, cleanliness and cost of production, and reduces reliance on skilled operators as compared with open processing.

5.2.6 Batch versus Continuous Separation

In view of cell manufacturing for cell therapy, which will have to adhere to Good Manufacturing Practice, it is important to consider whether a methodology can be implemented within a continuous process as opposed to a batch approach. In general, in bioprocessing, distinctions between batch and continuous downstream processes often lead to advantages one way or the other and are frequently related to the batch/continuous nature of the upstream production. A key advantage of batch processing, for example, is the ability to accurately distinguish the start and end of the production of a batch, which makes regulatory clarification significantly easier. On the other hand, batch processes frequently require "hold" steps to match processing times to equipment – with associated increases in sample handling and therefore risk of contamination with adventitious pathogens – and are therefore significantly harder to integrate into closed systems. Within the production of cellular therapies, however, these advantages either way may not be so clear-cut. It is still likely though that closed processing systems will offer significant production advantages, and thus a separation methodology that can be part of a continuous, closed separation process would likely be important for large-scale cell manufacturing.

5.3 Currently Developed Cell Separation Techniques

In this section we will list, in alphabetical order, reported cell separation techniques potentially suitable for large-scale clinical cell production. For each separation method we first describe the underlying technical principle, followed by discussion of the above listed selection criteria.

5.3.1 Acoustophoresis

Acoustophoretic sorting separates cells based on cell size, density and compressibility. It is a label-free, active method that employs an ultrasonic standing wave. The method combines a laminar flow with an axial acoustic primary radiation force in an ultrasonic stationary wave field (Figure 5.3). It makes use of the principle that cells subjected to sound waves experience a pressure gradient and travel laterally towards either the pressure node or antinode of the ultrasonic waves, depending on the cell's size and deformability. Although acoustophoresis employs an external force, it has been described as a gentle separation method with no apparent acute adverse effects for the cells due to exposure to ultrasound, including possible non-thermal effects due to cavitation (formation of jet streams due to collapsing gas bubbles) (Burguillos et al., 2013, and references therein). To prevent cell damage due to temperature elevation, separation could be performed in a controlled temperature environment (Burguillos et al., 2013).

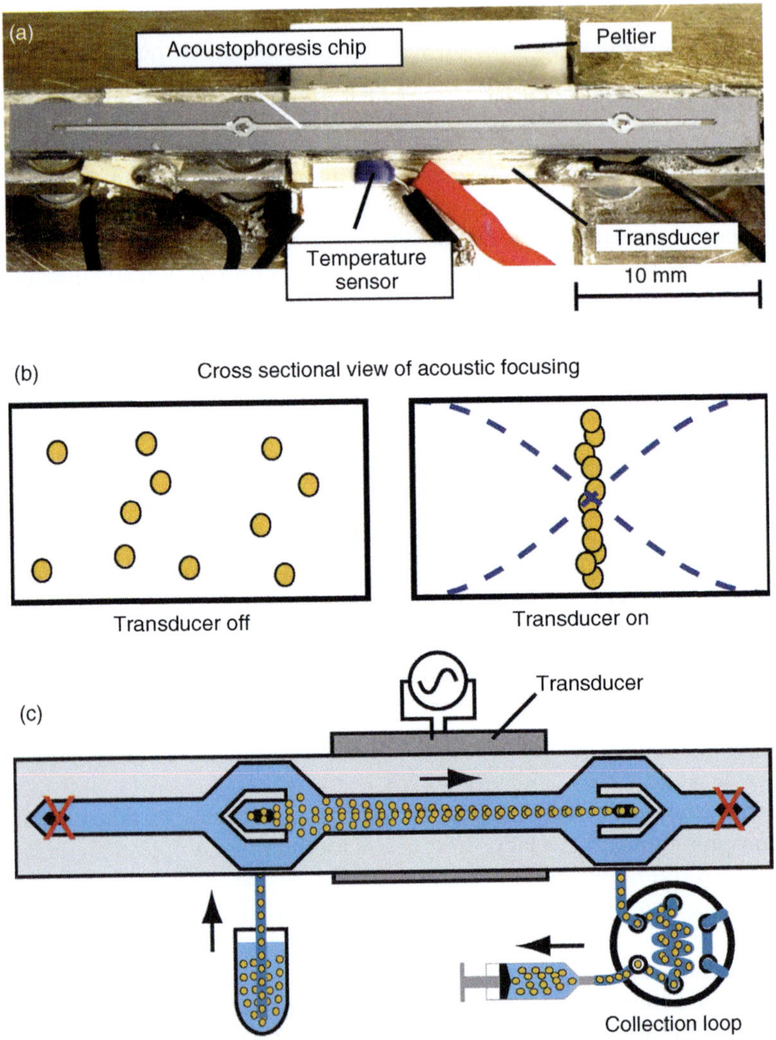

Figure 5.3 Principle of acoustophoresis: (a) Photo of the acoustophoresis microfluidic system; (b) Cross-sectional view of cell distribution in the microchannel without ultrasound (left) and with ultrasound forming an ultrasound standing wave (right); (c) Illustration of acoustophoresis chip (reprinted from Burguillos *et al.*, 2013).

The reported separation resolution is high, and can be optimised for required cell type by adjusting operating settings. To achieve separation of cell types with similar behaviours in an acoustic field force, acoustophoresis could be combined with medium-density manipulation by adjusting the density of the flow medium through addition of a compound like cesium chloride (Petersson et al., 2007). However, depending on downstream use, an additional medium wash step may be required following separation.

To date, reported acoustophoretic sorting devices have been developed as small, microfluidic platforms, not amenable to high throughput (in their current form) unless operated in parallel. Moreover, increasing sample flow rates in an attempt to increase throughput per device reportedly reduced recovery efficiency (Yang and Soh, 2012). Thus, for this separation technique to be implemented in a truly high-throughput manufacturing system, it will need to be developed further. Scalability is further hindered at present by the difficulties in increasing channel size for separation due to the need to locally focus the stationary wave. Increasing scale would also result in increased power consumption and heat generation, although these challenges could presumably be overcome by increased cooling or circulation. Its methodology, however, would allow it to be part of an integrated process, with continuous cell feeding and harvesting through syringe/pump operated systems in a closed platform. Furthermore, methods such as this that can loosely be grouped under the "field flow fractionation" (FFF) banner can potentially be scaled by moving from a 1D to a 2D or even 3D field environment, although this has proved more challenging with some fields than others.

5.3.2 Aqueous Two-Phase System (ATPS)

The Aqueous Two-Phase System (ATPS) separates cells according to size, hydrophobicity and net charge. The method, which is label-free and active, is a liquid-liquid fractionation method based on differential affinity of biological materials for either of two immiscible polymers. Commonly used polymers are polyethylene glycol (PEG) and dextran, with the more hydrophobic PEG forming an upper phase, which is of a lower density than the more hydrophilic dextran solution.

ATPS has been used for the enrichment of leukocytes within whole blood samples, achieving 9-fold enrichment (Soohoo and Walker,

2009), which is likely not efficient enough for cell therapy manufacture. Moreover, mammalian cell separation by ATPS is typically performed with a microfluidics device unsuitable for large-scale separation for cell manufacture (Nam et al., 2005; Tsukamoto et al., 2009), and development of macroscale ATPS for this application awaits further exploration. ATPS is routinely used in process engineering and industrial biotechnology for purification of smaller and less complex targets such as proteins however, and has been reasonably successfully operated at a large scale in these fields. A further challenge for cellular preparations is that the cells will be dissolved in polymer-containing medium, necessitating additional wash steps of the selected cells incorporated in the system to obtain a truly closed, integrated and continuous processing platform. In addition to label-free ATPS, immunoaffinity ATPS, using a monoclonal antibody to partition the target cell in one of the polymer phases, has been successfully applied to isolate cells since the 1980s (Karr et al., 1988), and has recently been reported to increase the CD34+ cell fraction in cord blood donations from 0.2% up to 42% (Sousa et al., 2011).

5.3.3 Centrifugal Techniques

Centrifugal separation of cells is not new, but the principle has been used in the design of novel cell processing methodologies like Centrifugal Counterflow Elutriation and integrated automated platforms combining centrifugation and filtration.

5.3.3.1 Centrifugal Counterflow Elutriation (CCE)

Centrifugal Counterflow Elutriation (CCE) separates cells based on their size and sedimentation density. It is a label-free, active method, employing lift forces. Cells in a conical centrifugal chamber are subjected to centrifugal forces in an outward direction and a fluid force in an inward direction. Since its introduction in the 1970s, CCE has been used routinely for research purposes, and later also for clinical procedures, including plasmapheresis and the collection of particular white blood cells and stem cells (Coulais et al. 2012; Dlubek et al., 2002; Faradji et al., 1991; Powell et al., 2009; Schouten et al. 1990; Tran et al., 2012) from apheresis products. Commercially available CCE platforms include the Beckman Coulter JE-5.0 Elutriation System, the Terumo BCT's closed system clinical grade Elutra© Cell Separation System and kSep® Systems centrifugation devices, kSep400 and kSep6000S. Within these platforms, CCE is already part of an automated, closed, integrated cell separation process capable of washing and then separating large batches of heterogeneous cell populations. For example, the capacity of the Elutra© processing chamber is 3×10^{10} white blood cells, while the kSep6000S is able to separate up to 1,200 billion cells at $10–50 \times 10^{6}$ cells/mL in a single use, fully closed, automated GMP-compatible platform. Whether CCE, and these commercially available systems in particular, lend themselves equally well for the separation of undifferentiated pluripotent stem cells from their *in vitro* differentiated progeny, needs to be determined.

Clearly centrifugation itself is a relatively easily scaled unit operation, and as such should present a strong case for incorporation in large-scale cell therapy purification. There are, however, two challenges to this. Firstly, CCE is not strictly a continuous centrifugation operation. The necessity of maintaining the density gradient within the separating chamber means that larger scale requires larger chambers, not increased continuous throughput, and eventually this becomes problematic from a rotational stability point of view. Furthermore, as the size of the device increases, physical shear forces on the cells during introduction increase, running the risk of damage to sensitive cell products.

5.3.3.2 Centrifuge Systems with Integrated Filters

Two commercially available automated centrifuge systems for blood component processing, the TACSI PL and the COBE 2991 Cell Processor, combine centrifugation with filtration (Terumo BCT). The TACSI PL (Terumo Automated Centrifuge and Separator Integration System for Platelets) is a closed, integrated manufacturing system that can process up to 6 buffy-coats at a time into platelet concentrate by first pooling and centrifugation of the buffy coats followed by pump-operated filtration for separation and leukoreduction of the final platelet product. Reports show that the system, which is used in blood centres, achieves efficient platelet recovery (Sandgren et al., 2010) suitable for clinical use (Lotens et al., 2014).

The COBE 2991 Cell Processor contains a centrifuge with a flexible diaphragm and separates components based on specific gravities. It is an integrated system whereby a hydraulic fluid presses against a flexible membrane to express fluid or cells into an attached collection bag. It is a versatile system that can be used for blood and cell sample washing, volume reduction of cell products and cell concentration. For specific cell selection purposes, for example human pancreatic islet isolation however, it does not seem to improve on manual density gradient purification (Shimoda et al., 2012).

In terms of scalability and engineering challenges, these systems are similar to the CCE approach described previously. However, it should be easier to develop a continuous version of a centrifuge/filter system, potentially simplifying scalability.

5.3.4 Dielectrophoresis (DEP)

Dielectrophoresis selects cells based on a variety of parameters, including size, shape, cytosol conductivity, polarisability and membrane properties like permeability, capacitance and conductivity. DEP-based separation is an active, label-free method, whereby an electric force is exerted on dielectric particles, including biological cells, suspended in a non-uniform electric field. The strength of this exerted force is dictated by properties of not only the cells but also the media, as well as the frequency of the electric field, and dictates whether a cell is attracted to or repelled by the charged electrode. Adjustment of the DEP force field can be obtained through an abundance of tunable factors, allowing manipulation of the cells according to various cell intrinsic properties. The technique has been used to separate a variety of mammalian cell types, including cancer cells (Becker et al., 1995), infected red blood cells (Valero et al., 2010) and undifferentiated pluripotent stem cells from their differentiated progeny (Flanagan et al., 2008; Velugotla et al., 2012).

Though DEP has been successfully integrated into a commercially available cell selection technology (i.e. ApoStream® available from ApoCell), this system is currently being used to select rare cells (e.g. circulating tumour cells) from a population, so distinct from separation in the context of a cell production processes. Indeed, a number of bottlenecks have been raised that likely prevent scalable DEP-based separation platforms for cell manufacture. Joule heating near the electrodes due to media conductivity may result in:

- bubbles and heat-induced cell death;
- redox reactions when DC current is utilised, with subsequent problems including the generation of free radicals;
- the strong dependency of DEP on the conductivity of the media used; and

- the strong dependency of DEP on cell size, such that inter cell-type size variation may override cell-type specific differences, and thus reduce resolution.

The latter hurdle, however, might be overcome with the recently developed isoelectric DEP, where sorting is mainly dependent on the dielectric properties rather than size by using a media conductivity gradient (Vahey and Voldman, 2008). Potentially, DEP could be part of a closed, continuous, cell separation platform that could achieve large-scale selection through scale-out rather than scale-up, by placing multiple devices in parallel. In terms of scale-up, DEP presents the same challenges as other FFF techniques, which potentially could be overcome by the development of 2D and 3D fields.

5.3.5 Deterministic Lateral Displacement (DLD)

Deterministic Lateral Displacement (DLD) is a label-free, passive technique that separates cells according to size and deformability. DLD is a continuous separation process based on asymmetric bifurcation of a laminar flow around obstacles (micropillars), with particles above a critical size being bumped laterally from the original streamline (Huang et al., 2004). Although DLD's simplicity is attractive, and DLD cell separation achieves good size resolution (Davis et al., 2006) and high enrichment (Zhang et al., 2012a), clogging of the device could be an issue for its implementation in an integrated, closed separation platform for cell manufacture. This is particularly likely to be the case in a sample containing a large ranges of cell sizes, as DLD devices tend to be inaccessible to particles significantly larger than their chosen separation diameter. However, reducing the pillars to point-like obstacles (thus reducing fluidic resistance) may improve throughput (Long et al., 2008). Furthermore, DLD is still attractive as a final "polishing" step, where contaminants may be low in number and similar in properties to target cells.

Potentially DLD could be scaled up, but there are manufacturing challenges in constructing large micropillar arrays. However, translation from 2D to 3D grids may again offer a solution.

5.3.6 Genetic Engineering

Genetic engineering has been used to select pure cultures of cardiomyocytes from differentiating embryonic stem cells (ESC) by the Field lab (Klug et al., 1996; Zandstra et al., 2003). This selection – which is not strictly a downstream technique like the others discussed in this chapter – was achieved by transfecting ESC with a fusion gene consisting of the alpha-cardiac myosin heavy chain (MHC) promoter driving the aminoglycoside phosphotransferase gene (neomycin resistance). These ESC were subsequently differentiated towards cardiomyocytes and enriched using G418 selection to generate more than 99% pure populations, though with varying yields. The authors claim that this approach could enable large-scale production of viable, ES cell-derived cardiomyocytes for tissue engineering and/or implantation, but as far as we are aware, to date no further studies have been published using genetic engineering for cell selection for clinical cell manufacture.

5.3.7 Hydrodynamic Filtration (HDF)

Hydrodynamic filtration (HDF) selects cells according to size and shape (Yamada and Seki, 2005). HDF is a label-free, active method whereby cells in a continuous flow are separated based on their flow profile as dictated by their size, which subsequently affects their proximity to the channel wall. HDF devices contain a main flow channel with a number of sideway collection outlets. Smaller cells, which will flow closer to the channel sidewall than larger cells, will enter the outlets earlier, and are thus the first to leave the channel. Separation is based purely on cell flow profile and not channel geometry. This means channels can be a lot larger than the cells, reducing clogging and increasing throughput, which is advantageous for scalability. The method has been used successfully to separate various cell types, including leukocytes (Yamada and Seki, 2005) and liver cells (Yamada et al., 2007) and, using HDF devices with curvilinear geometry to add Dean drag force, blood cells from plasma and WBC from RBC (Nivedita and Papautsky, 2013). HDF devices achieve high separation efficiency (ca. 95%; Nivedita and Papautsky, 2013) at high throughput rates (between 2×10^5 cells/min (Yamada et al., 2007) and 1×10^6 cells/min (Nivedita and Papautsky, 2013), but to date devices are small-scale and are being optimised for point-of-care applications rather than cell manufacturing (Nivedita and Papautsky, 2013).

5.3.8 Immunoadsorption

Immunoadsorption or affinity-based separation is a passive selection technique based on adherence of cells to molecules that are immobilised on a stationary matrix of some form (Figure 5.4). This matrix can be

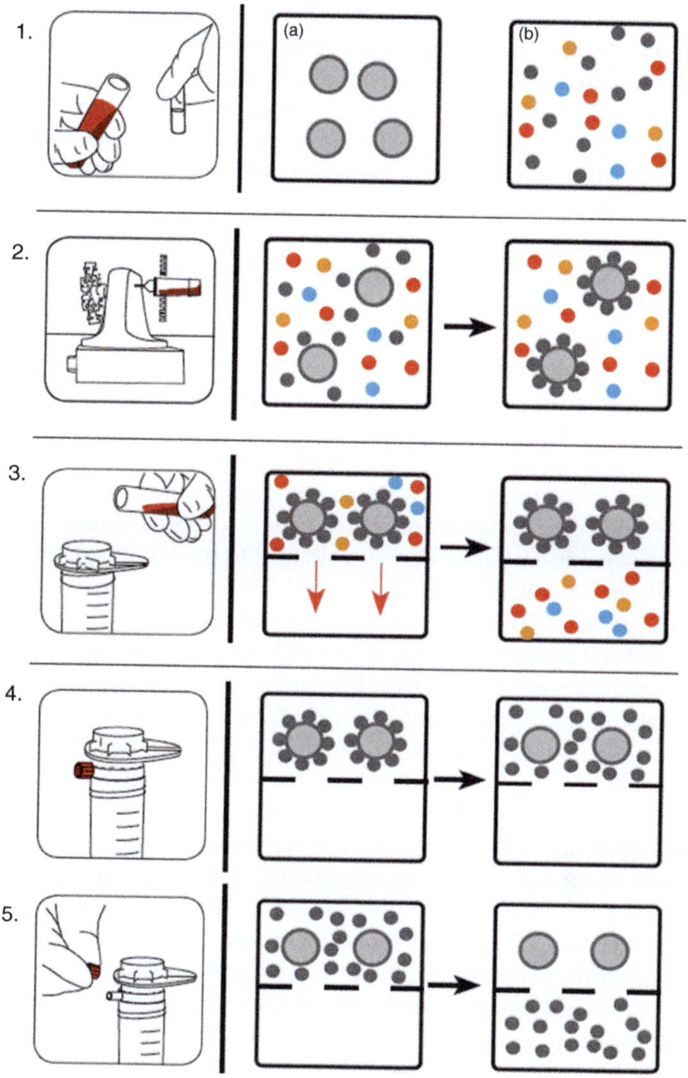

Figure 5.4 Principle of immunoadsorption. pluriBead® is a cell separation technology that works without magnetic components: (1) pluriBeads (a) are added to a heterogeneous cell sample (b); and (2) bind target cells; (3) Next, the sample is sieved through a strainer, after which the pluriBeads with bound target cells stay on top, while unwanted cells pass through; (4) Target cells are then detached using a proprietary detachment buffer; and (5) collected (reproduced with permission of pluriSelect).

presented on a variety of structures, including resin beads packed into columns or flasks, membranes, microposts and cryogels or monolithic columns. Various molecules can act as a ligand on the matrix, including proteins, peptides, aptamers and antibodies. Cells able to bind the ligand (the positive fraction) will be selectively adsorbed, while negative cells will flow through or remain in solution unhindered. If the captured cells are the cells of interest, they would then need to be detached using a matrix-specific method. This renders this technique more suitable for research and diagnostics (Didar and Tabrizian, 2010; Nagrath et al., 2007; Wang et al., 2007) than for the recovery of viable, unaltered cells for cell manufacture. Depending on the concentration of the target cell in the feed population, and the cell's affinity for the ligand, high purity (>97%) and sensitivity can be obtained with immunoadsorption (Cardoso et al., 1995; Dainiak et al., 2007; Murthy et al., 2004; Nagrath et al., 2007; Wang et al., 2007). Though most immunoadsorption platforms are designed as capture and release systems, the method can be integrated in a flow sorting approach to allow continuous separation based on cell rolling differences, albeit with low throughput (Karnik et al., 2008).

Recently, pluriSelect developed pluriBead, a combination of affinity-based cell selection and filtration, with antibody-mediated binding of cells to beads of varying sizes that can be separated from unlabelled cells with sieves. Pierzchalski et al. (2013) reported simultaneous separation of CD4+ and CD8+ cells from EDTA-whole blood using pluriSelect with high purity (95% and 90%, respectively) and viability (94% and 87%, respectively) in one hour. However, the current platform is unsuitable for large-scale cell separation, as it consists of cell strainers and tubes, and allows maximally 5×10^7 target cells per isolation, thus limiting its use for isolated batch-wise cell separation (Pierzchalski et al., 2013). It should be noted that, while immunoadsorption in the form of affinity chromatography is widely used in the large-scale commercial manufacture of protein therapeutics, that does not automatically make it a suitably scalable technique for cell targets. The specific challenges of applying chromatographic style separations to cell targets are discussed at length elsewhere (Willoughby, 2009), but are primarily driven by the partial incompatibility of the kinetics and diffusion limits of adsorption when coupled with the open flow paths necessary for cells to pass through a matrix. Specific capacity of the matrix is also very low for cells (in terms of number) when compared to proteins due to the relatively large size of cells.

5.3.9 Immunomagnetic Cell Sorting

Immunomagnetic cell separation devices make use of magnetic particles that carry specific antibodies to label target cells and require a magnetic field as the active force to collect cells tagged by the particles. Established immunomagnetic platforms, including Miltenyi's MACS system, rely on batch-wise cell capture and release, but other systems have been developed that allow continuous flow cell sorting. These include the dipole magnetic flow fractionator (DMF) and the quadrupole magnetic cell sorter (QMS) methods developed by Zborowski and Chalmers, based on split flow thin fractionation (SPLITT) (Chalmers et al., 1998; Nakamura et al., 2001; Sun et al., 1998). They consist of two and four magnetic poles, respectively, and fraction the feed population into multiple streams, depending on the cell's response to the magnetic field, which correlates with its label affinity (Figure 5.5). High-throughput cell separations with high purity and recovery have been reported for immunomagnetic continuous flow cell sorting, especially the QMS (e.g. 3.29×10^5 carcinoma cells/s, with 89% recovery (Nakamura et al., 2001)).

Another platform for continuous flow immunomagnetic cell sorting is BioCEP's Cell Enrichment Process (CEP), which combines flow-through technology with a unique "virtual mesh" system that holds the labelled cells contact free in stasis, while non-target cells flow out of the system. To date, no peer-reviewed studies have been published describing cell isolation using CEP. BioCEP's platform CEPir (CEP for isolation of rare cells) is a closed, integrated, research-use only device, which includes four electromagnets, a pump-controlled flow and software that allows definition of parameters like the intensity of the magnetic field, system capacity and separation time. CEPir can be operated in batch (0.5–50 ml), as well as dialysis mode with high volume/throughput (10^9 cells).

Large-scale operation of any immunomagnetic separation is limited by the need for larger magnetic fields. Similar developments in immunomagnetic chromatographic separations in the past have been restricted in size due to the heating side effects of the electromagnets necessary to generate fields in larger columns and systems.

5.3.10 Inertial Migration

Inertial migration is a size-based cell separation method. A label-free, active technique, it makes use of intrinsic hydrodynamic forces

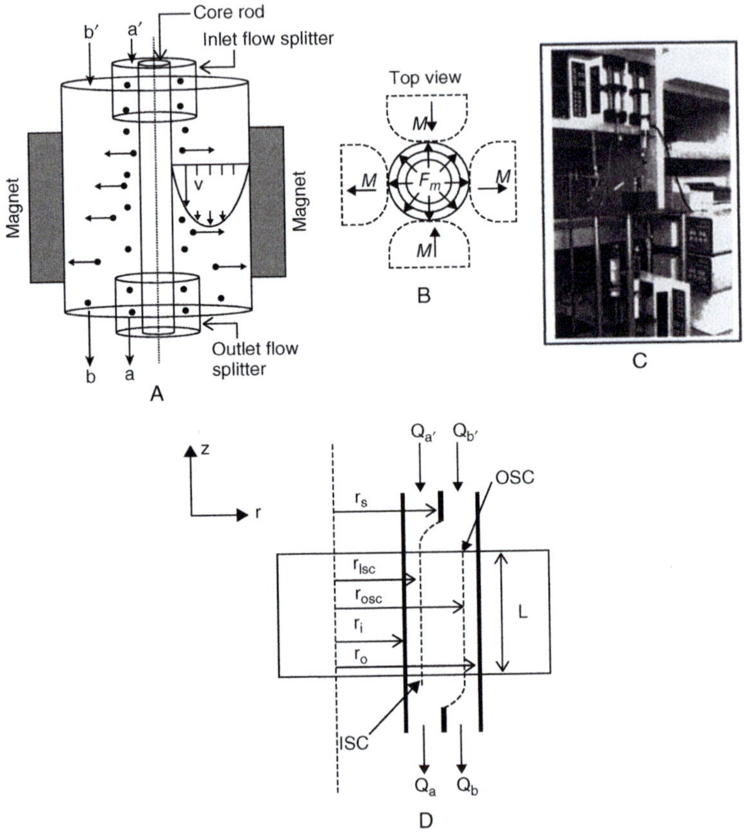

Figure 5.5 Principle of quadrupole magnetic cell sorting. Schematic drawing of the separation column of the Quadrupole magnetic cell sorter (QMS) showing cell injection port (a′), carrier (sheath fluid) injection port (b′), and elution of non-magnetically labelled cells (a) and magnetically labelled cells deflected by the magnetic field (b). The top view, B, shows the magnetic forces acting on the column to deflect cells. C is a photograph of the bench scale system. D is an enlarged diagram of the magnetic separation zone indicating the location of the inner and outer splitting cylinder (reproduced from Lara et al., 2006, Wiley Periodicals, Inc., copyright © 2006).

(Bhagat et al., 2008; Hur et al., 2011a, 2012; Kuntaegowdanahalli et al., 2009; Lee et al., 2013). Different types of inertial migration devices have been designed, consisting of straight, or one or more circular pipes or spiral channels (asymmetric curvilinear channel).

In all designs, cells are focused due to the balance of two inertial lift forces (known as the "tubular pinch effect"): i) shear gradient lift, and ii) wall effect lift. In the curvilinear-channel design, the Dean drag force is added as a third force, allowing multi-size separation. As a result, cells of a particular size are focused in a unique equilibrium position, generating distinct particle streams, which are collected in separate outlets at the end of the channel. Separation efficiency of this technique is inherently linked to cell population size heterogeneity, and reached for example 80% when used to separate two types of neuronal cells (neuroblasotoma and glioma cells) (Kuntaegowdanahalli et al., 2009). Due to its relatively simple design, the technique can be implemented as a closed, integrated platform for continuous cell selection. Similar to HDF, the channel width can be much larger than the cell sizes, allowing high volumetric flow rates ($>10^6$ cells/min), while reducing the chance of clogging.

Sim et al. (2011) have developed various multi-orifice flow fractionation (MOFF) devices for continuous cell separation, which utilise inertial lift and momentum-change-induced inertial force generated in a series of contraction/expansion microchannels, and show they can be used to efficiently select rare cells from blood samples (e.g. circulating tumour cells), but to date these devices have not been reported for use in large-scale cell separation (Moon et al., 2013; Sim et al., 2011).

5.3.11 Magnetic Cell Sorting – Label-free

Label-free magnetic cell sorting can be achieved with two approaches:

1) by using magnetophoresis; or
2) by embedding cells in a magnetic solution where the cells act as magnetic voids.

Both methods are active methodologies, requiring an external magnetic field, and to date have only been built as small-scale devices not suitable for large-scale use.

Potential bottlenecks for up-scaling the methods are possible attachment of cells to the channel wall and the effect of heat produced from large-scale electro-magnetic field generation on cell viability.

5.3.11.1 Magnetophoretic Cell Separation

Instruments for magnetophoretic cell separation distinguish cells based on differential magnetic susceptibility of cell (–content) and consist of a column placed in a magnetic field, for example a packed-column high-gradient magnetic separator (HGMS) (Jin et al., 2012). Magnetophoretic cell separation was first reported by Zborowski et al. (1995) for the selection of human lymphocytes labelled with ferritin and of red blood cells utilising the iron content of hemoglobin. Later applications include separation of red and white blood cells from diluted whole blood based on the cell's native magnetic properties (Han and Frazier, 2006), and selection of maturing RBC derived from hematopoietic stem cells through the separation of deoxygenated cells into paramagnetic cells (which stuck to the column) and dimagnetic cells (which flowed through the column), with the enriched population containing 80% of the target cells (Jin et al., 2012). Magnetophoretic selection has also been applied to separate cardiomyocytes, with up to 93% enrichment, from a digested rodent heart isolate (containing fibroblasts, smooth muscle cells and endothelial cells), after rendering the former transiently paramagnetic by oxidation of myoglobin into metmyoglobin by treating the cells with a NaNO3 solution (Sofla et al., 2013).

Thus far, reported yields are limited due to column capacity, but might be improved by parallel column placement, or design adaptations to develop continuous flow platforms, for example by using a U-shaped column (Liang and Xuan, 2012) or two electromagnets with anti-parallel current flow (Plouffe et al., 2012). Moreover, incorporating a deoxygenator in the platform will reduce sample handling steps required for the technology, which in its current form is unsuitable for closed, embedded cell manufacturing systems for the clinic.

5.3.11.2 Magnetic Solution-based Separation

The second label-free magnetic separation approach makes use of magnetic repulsion forces acting on cells (flowing through a microchannel) surrounded by a biocompatible magnetic fluid (Figure 5.6). This fluid will act as a uniform magnetic environment when exposed to an external magnetic field. The resulting effect is that the (largely non-magnetic) cells are pushed away, with bigger cells receiving a larger push force. By adjusting the magnetic susceptibility of the surrounding medium, the difference in the magnetic forces acting on different-sized cells can be altered. Cells are thus separated in various steams according to cell size. Fluids that have been used in this type of magnetic cell sorting are magnetic salt solutions (Shen et al., 2012) and ferrofluids (Kose et al., 2009). Reported resolution and purities

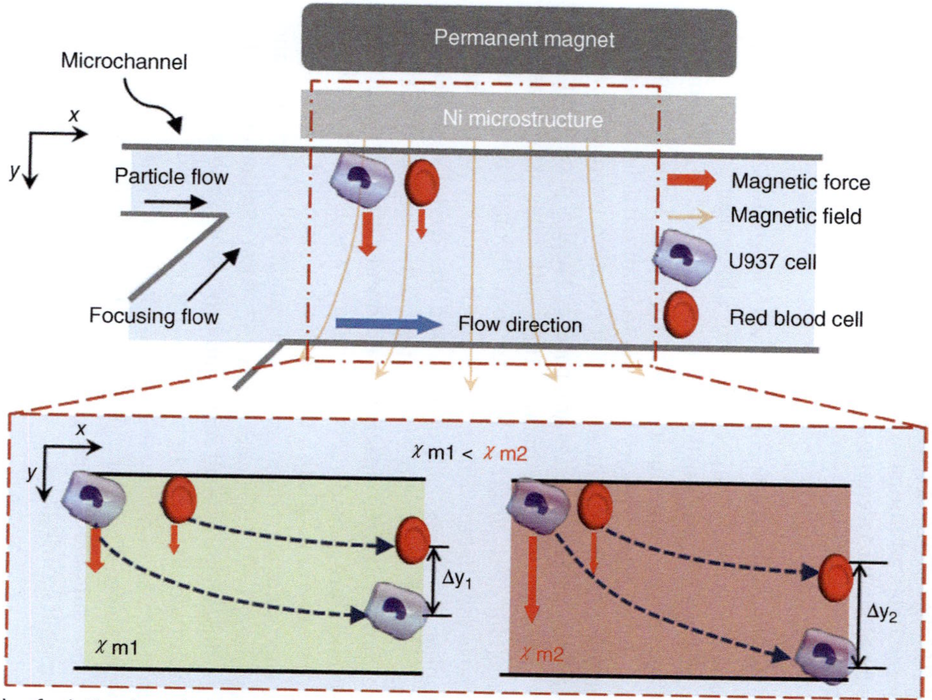

Figure 5.6 Principle of solution-based magnetophoretic separation. Schematic illustration of magnetophoretic cell separation using a magnetic repulsion force in a paramagnetic solution. When a magnetic field perpendicular to the direction of flow is applied by a permanent magnet placed next to the Ni microstructure, cells in a paramagnetic solution are deflected laterally due to the magnetic repulsion force. The separation efficiency is greatly enhanced by changing medium magnetic susceptibility from $\chi m1$ to $\chi m2$ via use of paramagnetic salt solutions (reprinted with permission from Shen *et al.* 2012, American Chemical Society, copyright © 2012).

vary, with Shen et al. achieving more than 90% purity when separating U937 cancer cells from RBC, while Kose et al. obtain a separation efficiency of 75% and purity of 89% when separating sickle cells from healthy blood cells. Though label-free, the method is invasive, as the cells need to be suspended in a magnetic medium, the impact of which on downstream clinical applications remains to be determined, and which necessitates additional wash steps, complicating incorporation of the technology into a continuous, closed, embedded process.

5.3.12 Microscale Vortices

Microscale laminar vortices combined with inertial focusing enables size-selective trapping of larger cells from a heterogeneous cell population, while flushing out smaller cells (Hur et al., 2011b). The method is label-free and does not require an external force. Instead, it employs lift forces to establish irreversible migration of particles into microscale vortices that develop in an array of expansion-contraction trapping reservoirs placed in series and parallel. Parallel placement allows for high throughput device manufacture, with rates as high as 7.5×10^6 cells/s (Hur et al., 2011b). However, total throughput may be limited because the current design is not continuous due to the fact that the trapping reservoirs lack outlets, and the device operates with vortex-priming, flush and releasing steps. Although the Di Carlo lab initially reported that the enrichment ratio and capturing efficiency of their method were considerably lower than various established separation methods (Hur, 2011b), they recently optimised the design to achieve successful extraction and enumeration of CTC from a cancer patient's blood with improved purity (57–94%) (Sollier et al., 2014). However, the design is small scale, and processing time is 20 minutes for 7.5 ml of whole blood. In addition, the vortices will likely create gas–liquid interfaces during separation, which is a particularly harmful type of shear stress for cells. Taken together, the methodology seems more suitable for cell selection for diagnostics than for cell therapy manufacturing.

5.3.13 Normal Flow Filter (NFF)

Normal flow filter devices (also called microfilters) contain small filters (sieves) with circular or oval holes to selectively capture cells based on size and deformability. Since these devices are not reliant on an external force or cell labelling, they can be used as a simple dead-end filter, trapping unwanted cells. As a consequence, the method works best if the percentage of unwanted cells is relatively low, so as not to clog the filter too quickly. In addition, clogging of the pores may result in flow irregularities. The platform has been used for research and diagnostics in the cell therapy area (e.g. trapping of rare circulating cancer cells; Zheng et al., 2007, 2011) but not for (large-scale) cell production for the clinic. However, NFF for cell removal is routinely used at very large scale in the manufacture of protein therapeutics from mammalian cells, so clearly there is scope for NFF to work as a scalable technique.

To prevent clogging and allow continuous separation, Sethu et al, (2006) arranged the sieves on the side of the flow channels and connected the channel to a flare-shaped diffuser. At the right flow rate (i.e. pressure gradient), this diffuser acts as the outlet of the fraction enriched for the larger cells that were unable to pass through the sieves due to size and deformability constraints. Fractionating peripheral blood with their device at a flow rate of 5 ul/min, they achieved close to 99% leukocyte depletion and ca. 50% isolation of total erythrocytes. Erythrocyte retrieval rates could be enhanced by increasing the flow rate, but this reduced selectivity (Sethu et al., 2006), and may increase shear stress and thus cell damage.

In general, NFF would only be suitable for negative selection processes such as the removal of larger cells from smaller target cells, but presents an interesting opportunity in some processes, particularly as a more crude early-stage separation technique.

5.3.14 Optical – Label-free

Optical sorting of cells is an active method selecting cells based on their size and refractive index using an optical force. Depending on the type of optical sorting method, cell labelling is essential, or optional, merely enhancing selectivity. Dholakia et al. (2007) call the former active, and the latter passive optical sorting. An example of active optical sorting is FACS, while passive optical sorting can be achieved with: i) optical tweezers (Ashkin et al., 1987; Chiou et al., 2005); ii) optical chromatography (Imasaka et al., 1995); or iii) optical energy landscapes (MacDonald et al., 2003, 2004). Optical tweezers use radiation pressure from a focused laser beam to trap and move cells (Figures 5.7 and 5.8), while optical chromatography employs the balance of fluidic and optical forces on a particle in a channel with a flow of fluid in the opposite direction to a laser beam (Figure 5.9).

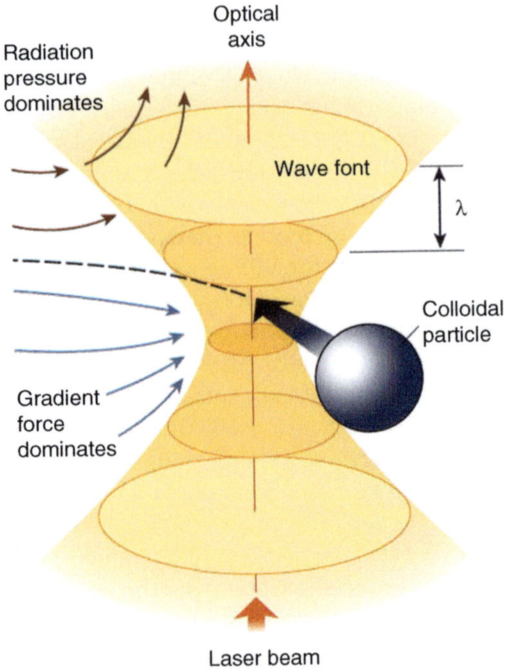

Figure 5.7 Principle of optical tweezers. Optical tweezers use a strongly focused beam of light to trap objects. Intensity gradients in the converging beam draw small objects, such as a colloidal particle, towards the focus, whereas the radiation pressure of the beam tends to blow them down the optical axis. Under conditions where the gradient force dominates, a particle can be trapped, in 3D, near the focal point (Grier, 2003, reprinted by permission from Macmillan Publishers Ltd, copyright © 2003).

Optical energy landscapes use the interference of two or more laser beams to generate an energy landscape, and utilise variation in shape, size and refractive index to induce separation movement in different populations within the landscape (Figure 5.10). Though optical sorting has been used successfully to separate distinct mammalian cell populations (Applegate et al., 2004; MacDonald et al., 2003, 2004), this can only be achieved when the differences in size and refractive index of the various populations is large enough (Paterson et al., 2007). Moreover, separation efficiency might be compromised due to the fact that a cell could have different refractive indexes depending on

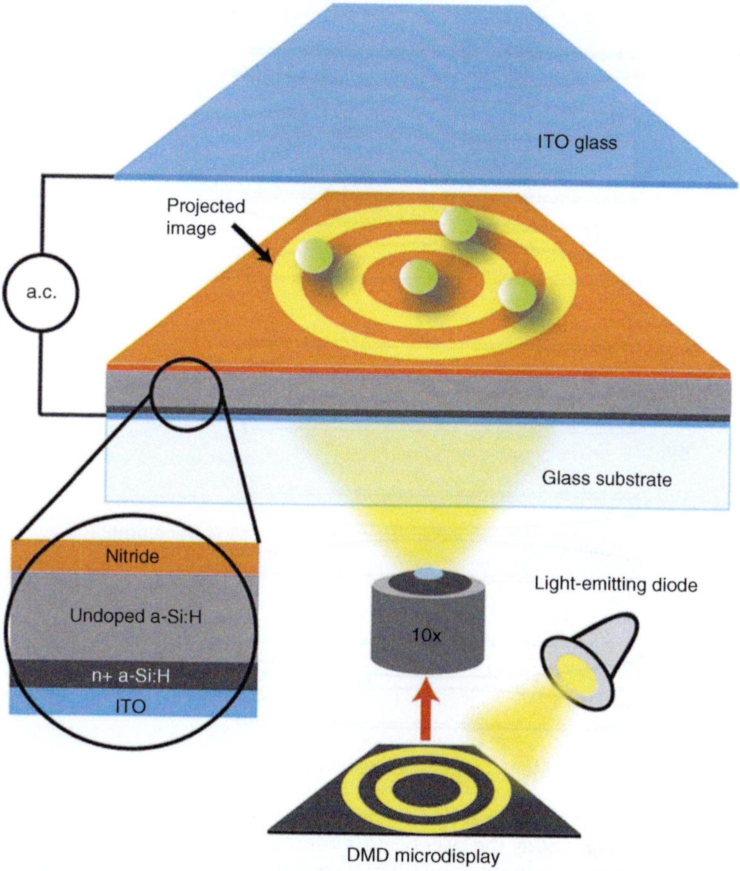

Figure 5.8 Principle of optoelectronic tweezers. Device structure used in optoelectronic tweezers. Liquid that contains microscopic particles is sandwiched between the top indium tin oxide (ITO) glass and the bottom photosensitive surface consisting of ITO-coated glass topped with multiple featureless layers: 50 nm of heavily doped hydrogenated amorphous silicon (a-Si:H), 1 mm of undoped a-Si:H and 20 nm of silicon nitride. The top and bottom surfaces are biased with an a.c. electric signal. The illumination source is a light-emitting diode operating at a wavelength of 625 nm (Lumileds, Luxeon Star/O). The optical images shown on the digital micromirror display (DMD) are focused onto the photosensitive surface and create the non-uniform electric field for DEP manipulation (Chiou et al., 2005, reprinted by permission from Macmillan Publishers Ltd, copyright © 2005).

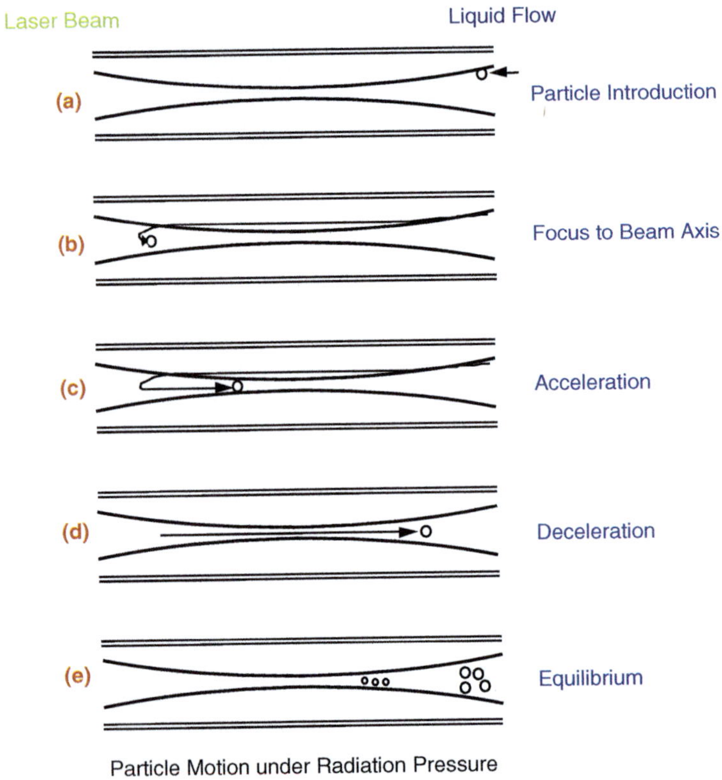

Figure 5.9 Principle of optical chromatography. Schematic of particle motion under radiation pressure. The laser beam is introduced from the left-hand side and the liquid from the right-hand side: (a) particle introduction; (b) focus of particle into beam centre by gradient force; (c) acceleration of particle; (d) deceleration of particle; and (e) particles drifting at equilibrium position, resulting in particle separation. Particle elution is then achieved by decreasing the laser power (smallest particles elute first, followed by subsequent particles upon further decreasing the laser power (Imasaka et al., 1995, reprinted (adapted) with permission from American Chemical Society, copyright © 1995).

the position or state of its various components. To date, optical separation in the literature has been achieved only with small-scale devices. However, the development of larger-scale and 3D optical landscapes is a focus of current research, and this approach may render the methodology suitable for adaption to larger-scale separations, particularly if power consumption can be addressed.

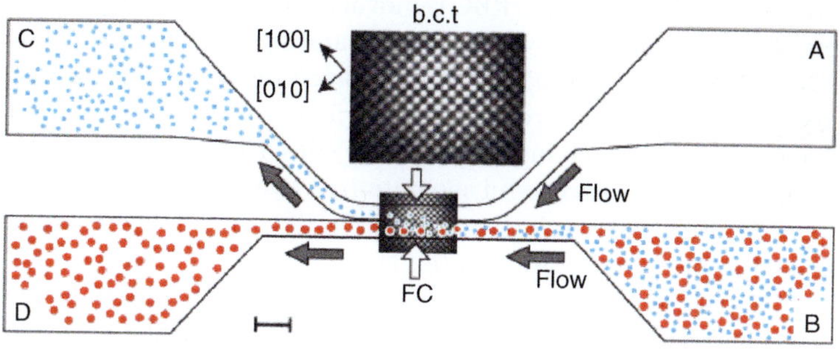

Figure 5.10 Principle of optical landscape fractionation. Low Reynolds number flows will be laminar: without an actuator all particles from chamber B would flow into chamber D. Chamber A would typically introduce a "blank" flow stream, although this could be any stream into which the selected particles are to be introduced. By introducing a 3D optical lattice – in this case a body-centred tetragonal (b.c.t.) lattice – into the fractionation chamber (FC), one species of particle is selectively pushed into the upper flow field. The reconfigurability of the optical lattice allows for dynamic updating of selection criteria. For weakly segregated species, the analyte can be either recirculated through the optical lattice or directed through cascaded separation chambers. This latter option also allows the use of multiple selection criteria in a single integrated chip (MacDonald et al., 2003, reprinted by permission from Macmillan Publishers Ltd, copyright © 2003).

5.3.15 Tangential Flow Filters (TFF)

Tangential Flow Filters (also known as cross-flow filters or microscale membranes) passively select cells based on size and especially deformability during laminar flow migration through a micropost array. The method does not require cell labelling. The name derives from the path the fluid follows in this flow, namely tangentially across the surface of the filter rather than into the filter (as is the case with the Normal Flow Filter). Subsequently, larger cells that may clog the filter can be washed away during the filtration process, delaying problems with obstructed flow. This in turn will increase the length of time the filter can be functional. Furthermore, cells or particles retained in a TFF system (the retentate) remain in solution/media and hence are more likely to remain healthy. Thus TFF is usable as both a positive and negative separation tool, where NFF would most likely only be useable for negative selection.

The technique has been used to isolate and enumerate cancer cells from blood (isolation efficiency >80% (Tan et al., 2009, 2010)) and blood separation (e.g. >95% RBC removal (Chen et al., 2008) and more than 97% plasma selectivity when separating plasma from whole blood (Chen et al., 2009)), but the reported devices are microscale, and samples require rigorous dilution (>50 times (Chen et al., 2008)) prior to separation. Zhang et al. (2012b) describe in their proof-of-principle paper the successful separation of two cancer cells with different deformability using a TFF chip device, but with limited efficiency (the fraction of the more elastic cells increased from 50 to 73%). A study comparing various silicon-based microfilters to separate white blood cells from red blood cells indicates that TFF is superior over NFF and weir and pillar type filters (Ji et al., 2008). TFF systems are commonly operated at large scale in the mammalian and bacterial cell culture industries and several companies produce large-scale TFF membranes suitable for handling mammalian cell material. However, care must be taken during scale-up to ensure that shear levels within the flow path of the membrane systems remain low enough to not adversely affect cells as, in general, recovery of healthy cells is not a priority in industrial mammalian cell TFF.

5.3.16 Weir and Pillar

Two other label-free filter methods that separate cells by forcing cells to flow past obstacles are: i) weir-type filters; and ii) pillar-type filters. Like TFF, both weir-and pillar-type filters passively separate cells based on size and deformability, but unlike TFF, the latter two usually employ a laminar flow that runs parallel to the flow channel rather than tangential. Weir-type filters, named after a type of dam with a similar design, exclude cellular components by allowing only smaller and/or more flexible cells to pass through a planar slit. Pillar-type filters achieve exclusion of larger and stiffer cells by adjusting the spacing between pillars. Both weir-and pillar-type filters have been reported as part of a continuous flow device (VanDelinder and Groisman, 2007).

Though the design simplicity is an attractive feature of these filters, they require design variations, adjusting slit- and gap sizes for different cell samples. In addition, clogging and fouling may be an issue as cells are trapped in the direction of the cell flow, rendering their use for large-scale cell production unlikely unless they are combined with other selection technique(s) and have adapted flow types/direction (e.g. pulsatile pressure (VanDelinder and Groisman, 2006)) or flow parallel to the weir structures (Chen et al., 2008). So far studies on fractionating whole blood report low flow rates and limited efficiency and a study by Ji et al. (2008), comparing TFF, NFF, weir-type filters and pillar-type filters for the separation of white blood cells from red blood cells showed TFF as being more efficient that the latter three. Weir-and pillar-type filters may thus be more suited for small-scale applications, for example obtaining a relatively low number of particular cells for downstream analysis rather than large-scale cell production processes.

5.4 Conclusion

To bring cell-based therapies from bench to clinic, cost-effective, scalable cell separation devices with high resolution are required for integration into GMP-compatible bioprocessing platforms. Whilst this is the case for cellular therapies in general, it must be considered a particular priority for allogeneic therapies (such as blood-related products) likely to be manufactured on a larger scale for multiple recipients. Though various new cell selection methodologies described in this chapter are promising candidates that can achieve high purity and selectivity, at the current level of development trade-off must usually be accepted in the form of low through-put, and most are therefore more suitable as point-of-care devices until further scalability has been addressed. Indeed, to date none can be seen to clearly perform better in terms of combined selectivity and throughput than currently available commercial systems like FACS and MACS. However, these established methods rely on sample labelling, rendering them less suitable for clinical applications. Moreover, due to their batch-type operation they do not lend themselves to be easily incorporated in integrated cell manufacturing systems and may be limited in terms of scale of operation. As far as we are aware, the centrifuge-based cell processing platforms described in this chapter (JE 5.0 Elutriation System (Beckman Coulter), kSEP (kSEP® Systems) and Elutra©, TACSI PL and COBE 2991 (Terumo BCT)) are the only label-free commercial cell handling devices that have demonstrated processing of clinical grade cellular/ blood-derived products in closed systems. The current “gold standard” of cell separation in clinical and production environments, however, are the Miltenyi MACS systems such as the CliniMACS Prodigy®, but these are label-based systems.

Table 5.1 The various cell separation methodologies set against 4 parameters, and scored for cell manufacturing potential based on type of production (batch or continuous) and system (closed or open)

Method	Label	Force	Production	System	Potential
Acoustophoretic	Free	Active	Continuous	Closed	++
ATPS	Free	Active	Continuous	Closed	++
CCE	Free	Active	Batch	Closed	+
Filter centrifugation	Free	Active	Batch	Closed	+
DEP	Free	Active	Continuous	Closed	++
DLD	Free	Passive	Continuous	Closed	++
Genetic	Free	Passive	Continuous	Open	-
HDF	Free	Active	Continuous	Open	-
Immunoadsorption	Partial*	Passive	Batch	Open	- -
Immunomagnetic	Partial*	Active	Continuous**	Closed	++
Inertial	Free	Active	Continuous	Closed	++
Magnetophoretic	Free	Active	Batch	Open	- -
Magnetic medium	Free	Active	Continuous	Open	-
Microscale vortices	Free	Active	Continuous	Open	-
NFF	Free	Passive	Continuous	Closed	++
Optical	Free	Active	Continuous	Open	-
TFF	Free	Passive	Continuous	Closed	++
Weir and Pillar	Free	Passive	Continuous	Open	-

* Label/antibodies are attached to stationary phase/binding adsorbent, not target, but may still end up in target population.

** Some variants of the technique can be considered semi-continuous.

Table 5.1 contains a summary of key selection parameters of cell separation methods described in this chapter, plus an indication of a method's potential, that is, its suitability for incorporation into integrated cell manufacturing platforms. Listed methods will require further bioengineering development and need to evolve from their current small-scale chip design to devices capable of high throughput to overcome the rate-limiting step – efficient cell selection – of bringing regenerative medicine, and especially stem-cell based therapeutics, to the patient. Given the number of methodologies under investigation, and the "usual" requirement in downstream bioprocessing to operate with multiple distinct separation steps, development of hybrid devices, combining advantageous aspects of various methodologies, is likely to prove a useful approach.

Acknowledgements

Dr Hoeve is a recipient of a Flexible Interchange Programme Award (BBSRC grant no. BB/L004925/1).

References

Applegate, R. Jr., Squier, J., Vestad, T. *et al.* (2004) Optical trapping, manipulation, and sorting of cells and colloids in microfluidic systems with diode laser bars. *Opt. Express*, **12** (19), 4390–8.

Ashkin, A., Dziedzic, J.M. and Yamane, T. (1987) Optical trapping and manipulation of single cells using infrared laser beams. *Nature*, **330** (6150), 769–71.

Becker, F.F., Wang, X.B., Huang, Y. *et al.* (1995) Separation of human breast cancer cells from blood by differential dielectric affinity. *Proc. Natl. Acad. Sci. USA*, **92** (3), 860–4.

Bhagat, A.A., Kuntaegowdanahalli, S.S. and Papautsky, I. (2008) Continuous particle separation in spiral microchannels using Dean flows and differential migration. *Lab Chip*, **8** (11), 1906–14.

Burguillos, M.A., Magnusson, C., Nordin, M., Lenshof, A., Augustsson, P. et al. 2013. Microchannel acoustophoresis does not impact survival or function of microglia, leukocytes or tumor cells. *PLoS One*, **8**(5), e64233.

Cardoso, A.A., Watt, S.M., Batard, P. *et al.* (1995) An improved panning technique for the selection of CD34+ human bone marrow hematopoietic cells with high recovery of early progenitors. *Exp. Hematol.*, **23** (5), 407–12.

Chalmers, J.J., Zborowski, M., Sun, L. and Moore, L. (1998) Flow through, immunomagnetic cell separation. *Biotechnol. Prog.*, **14** (1), 141–814.

Chen, X., Cui, D., Liu, C. and Li, H. (2008) Microfluidic chip for blood cell separation and collection based on crossflow filtration. *Sensors and Actuators B: Chemical*, **130**, 216–21.

Chen, X., Cui, D. and Zhang, L. (2009) Isolation of plasma from whole blood using a microfluidic chip in a continuous cross-flow. *Chin. Sci. Bull.*, **54**, 324–7.

Chiou, P.Y., Ohta, A.T. and Wu, M.C. (2005) Massively parallel manipulation of single cells and microparticles using optical images. *Nature*, **436**, 370–2.

Coulais, D., Panterne, C., Fonteneau, J.F. and Grégoire, M. (2012) Purification of circulating plasmacytoid dendritic cells using counterflow centrifugal elutriation and immunomagnetic beads. *Cytotherapy*, **14** (7), 887–96.

Dainiak, M.B., Galaev, I.Y., Kumar, A. *et al.* (2007) Chromatography of living cells using supermacroporous hydrogels, cryogels. *Adv. Biochem. Eng. Biotechnol.*, **106**, 101–27.

Davis, J.A., Inglis, D.W., Morton, K.J. *et al.* (2006) Deterministic hydrodynamics: taking blood apart. *Proc. Natl. Acad. Sci. USA*, **103** (40), 14779–884.

Dholakia, K., MacDonald, M.P., Zemánek, P. and Cizmár, T. (2007) Cellular and colloidal separation using optical forces. *Methods Cell Biol.*, **82**, 467–95.

Didar, T.F. and Tabrizian, M. (2010) Adhesion-based detection, sorting and enrichment of cells in microfluidic Lab-on-Chip devices. *Lab Chip*, **10** (22), 3043–53.

Dlubek, D., Dybko, J., Wysoczanska, B. *et al.* (2002) Enrichment of normal progenitors in counter-flow centrifugal elutriation (CCE) fractions of fresh chronic myeloid leukemia leukapheresis products. *Eur. J. Haematol.*, **68** (5), 281–8.

Faradji, A., Bohbot, A., Schmitt-Goguel, M. *et al.* (1991) Apheresis-elutriation program for adoptive immunotherapy with autologous activated monocytes in cancer patients. *Int. J. Artif. Organs*, **14** (5), 304–12.

Flanagan, L.A., Lu, J., Wang, L. *et al.* (2008) Unique dielectric properties distinguish stem cells and their differentiated progeny. *Stem Cells*, **26** (3), 656–65.

Grier, D.G. (2003) A Revolution in Optical Manipulation. *Nature*, **424**, 810–16.

Han, K.H. and Frazier, A.B. (2006) Paramagnetic capture mode magnetophoretic microseparater for high efficiency blood cell separations. *Lab Chip*, **2**, 265–73.

Huang, L.R., Cox, E.C., Austin, R.H. and Sturm, J.C. (2004) Continuous particle separation through deterministic lateral displacement. *Science*, **304** (5673), 987–90.

Hur, S.C., Henderson-MacLennan, N.K., McCabe, E.R. and Di Carlo, D. (2011a) Deformability-based cell classification and enrichment using inertial microfluidics. *Lab Chip*, **11** (5), 912–20.

Hur, S.C., Mach, A.J. and Di Carlo, D. (2011b) High-throughput size-based rare cell enrichment using microscale vortices. *Biomicrofluidic*, **5** (2), 22206.

Hur, S.C., Brinckerhoff, T.Z., Walthers, C.M., Dunn, J.C. and Di Carlo, D. 2012. Label-free enrichment of adrenal cortical progenitor cells using inertial microfluidics. *PLoS One*, **7**(10), e46550.

Imasaka, T., Kawabata, Y., Kaneta, T. and Ishidzu, Y. (1995) Optical chromatography. *Anal. Chem.*, **67** (11), 1763–5.

Ji, H.M., Samper, V., Chen, Y. *et al.* (2008) Silicon-based microfilters for whole blood cell separation. *Biomed. Microdevices*, **10** (2), 251–7.

Jin, X., Abbot, S., Zhang, X., Kang, L., Voskinarian-Berse, V., Zhao, R. et al. 2012. Erythrocyte enrichment in hematopoietic progenitor cell cultures based on magnetic susceptibility of the hemoglobin. *PLoS One*, **7**(8), e39491.

Karnik, R., Hong, S., Zhang, H. *et al.* (2008) Nanomechanical control of cell rolling in two dimensions through surface patterning of receptors. *Nano Lett.*, **8** (4), 1153–8.

Karr, L.J., Van Alstine, J.M., Snyder, R.S. *et al.* (1988) Cell separation by immunoaffinity partitioning with polyethylene glycol-modified protein A in aqueous polymer two-phase systems. *J. Chromatogr.*, **442**, 219–27.

Klug, M.G., Soonpaa, M.H., Koh, G.Y. and Field, L.J. (1996) Genetically selected cardiomyocytes from differentiating embryonic stem cells form stable intracardiac grafts. *J. Clin. Invest.*, **98** (1), 216–24.

Kose, A.R., Fischer, B., Mao, L. and Koser, H. (2009) Label-free cellular manipulation and sorting via biocompatible ferrofluids. *Proc. Natl. Acad. Sci. USA*, **106** (51), 21478–83.

Kuntaegowdanahalli, S.S., Bhagat, A.A., Kumar, G. and Papautsky, I. (2009) Inertial microfluidics for continuous particle separation in spiral microchannels. *Lab Chip*, **9** (20), 2973–80.

Lara, O., Tong, X., Zborowski, M. *et al.* (2006) Comparison of two immunomagnetic separation technologies to deplete T-cells from human blood samples. *Biotechnology & Bioengineering*, **94** (1), 66–80.

Lee, M.G., Shin, J.H., Bae, C.Y. *et al.* (2013) Label-free cancer cell separation from human whole blood using inertial microfluidics at low shear stress. *Anal. Chem.*, **2:85** (13), 6213–8.

Liang, L. and Xuan, X. (2012) Continuous sheath-free magnetic separation of particles in a U-shaped microchannel. *Biomicrofluidics*, **6** (4), 44106.

Long, B.R., Heller, M., Beech, J.P. *et al.* (2008) Multidirectional sorting modes in deterministic lateral displacement devices. *Phys. Rev. E.*, **78**, 046304.

Lotens, A., Najdovski, T., Cellier, N. *et al.* (2014) New approach to "top-and-bottom" whole blood separation using the multiunit TACSI WB system: quality of blood components. *Vox Sang.*, **107** (3), 261–8.

MacDonald, M.P., Spalding, G.C. and Dholakia, K. (2003) Microfluidic sorting in an optical lattice. *Nature*, **426** (6965), 421–4.

MacDonald, M.P., Neale, S., Paterson, L. *et al.* (2004) Cell cytometry with a light touch: Sorting microscopic matter with an optical lattice. *J. Biol. Regul. Homeost. Agents*, **18** (2), 200–5.

Moon, H.S., Kwon, K., Hyun, K.A. *et al.* (2013) Continual collection and re-separation of circulating tumor cells from blood using multi-stage multi-orifice flow fractionation. *Biomicrofluidics*, 7 (1), 14105.

Murthy, S.K., Sin, A., Tompkins, R.G. and Toner, M. (2004) Effect of flow and surface conditions on human lymphocyte isolation using microfluidic chambers. *Langmuir*, **20** (26), 11649–55.

Nagrath, S., Sequist, L.V., Maheswaran, S. *et al.* (2007) Isolation of rare circulating tumour cells in cancer patients by microchip technology. *Nature*, **450** (7173), 1235–9.

Nakamura, M., Decker, K., Chosy, J. *et al.* (2001) Separation of a breast cancer cell line from human blood using a quadrupole magnetic flow sorter. *Biotechnol. Prog.*, **17** (6), 1145–55.

Nam, K.H., Chang, W.J., Hong, H. *et al.* (2005) Continuous-flow fractionation of animal cells in microfluidic device using aqueous two-phase extraction. *Biomed. Microdevices*, 7 (3), 189–95.

Nivedita, N. and Papautsky, I. (2013) Continuous separation of blood cells in spiral microfluidic devices. *Biomicrofluidics*, 7 (5), 54101.

Paterson, L., Papagiakoumou, E., Milne, G. *et al.* (2007) Passive optical separation within a "nondiffracting" light beam. *J. Biomed. Opt.*, **12** (5), 054017.

Petersson, F., Aberg, L., Swärd-Nilsson, A.M. and Laurell, T. (2007) Free flow acoustophoresis: Microfluidic-based mode of particle and cell separation. *Anal. Chem.*, **79** (14), 5117–23.

Pierzchalski, A., Mittag, A., Bocsi, J. and Tarnok, A. 2013. An innovative cascade system for simultaneous separation of multiple cell types. *PLoS One*, **8**(9), e74745.

Plouffe, B.P., Lewis, L.H. and Murthy, S.K. (2012) Computational design optimization for microfluidic magnetophoresis. *Biomicrofluidics*, **5**, 013413.

Powell, D.J. Jr., Brennan, A.L., Zheng, Z. *et al.* (2009) Efficient clinical-scale enrichment of lymphocytes for use in adoptive immunotherapy using a modified counterflow centrifugal elutriation program. *Cytotherapy*, **11** (7), 923–35.

Sandgren, P., Hild, M., Sjödin, A. and Gulliksson, H. (2010) Storage of Buffy-coat-derived platelets in additive solutions: *In vitro* effects on platelets prepared by the novel TACSI system and stored in plastic containers with different gas permeability. *Vox Sang*, **99** (4), 341–7.

Schouten, H.C., Kessinger, A., Smith, D.M. *et al.* (1990) Counterflow centrifugation apheresis for the collection of autologous peripheral blood stem cells from patients with malignancies: A comparison with a standard centrifugation apheresis procedure. *J. Clin. Apher.*, **5** (3), 40–144.

Schwartz, S.D., Hubschman, J.P., Heilwell, G. *et al.* (2012) Embryonic stem cell trials for macular degeneration: a preliminary report. *Lancet*, **379** (9817), 713–20.

Sethu, P., Sin, A. and Toner, M. (2006) Microfluidic diffusive filter for apheresis (leukapheresis). *Lab Chip*, **6** (1), 83–9.

Shimoda, M., Itoh, T., Iwahashi, S. *et al.* (2012) An effective purification method using large bottles for human pancreatic islet isolation. *Islets*, **4** (6), 398–404.

Shen, F., Hwang, H., Hahn, Y.K. and Park, J.K. (2012) Label-free cell separation using a tuneable magnetophoretic repulsion force. *Anal. Chem.*, **3:84** (7), 3075–81.

Sim, T.S., Kwon, K., Park, J.C. *et al.* (2011) Multistage-multiorifice flow fractionation (MS-MOFF): continuous size-based separation of microspheres using multiple series of contraction/expansion microchannels. *Lab Chip*, **11** (1), 93–9.

Sofla, A., Cirkovic, B., Hsieh, A. *et al.* (2013) Enrichment of live unlabelled cardiomyocytes from heterogeneous cell populations using manipulation of cell settling velocity by magnetic field. *Biomicrofluidics*, **13:7** (1), 14110.

Sollier, E., Go, D.E., Che, J. *et al.* (2014) Size-selective collection of circulating tumor cells using Vortex technology. *Lab Chip*, **7:14** (1), 63–77.

Soohoo, J.R. and Walker, G.M. (2009) Microfluidic aqueous two phase system for leukocyte concentration from whole blood. *Biomed. Microdevices*, **11** (2), 323–9.

Sousa, A.F., Andrade, P.Z., Pirzgalska, R.M. *et al.* (2011) A novel method for human hematopoietic stem/progenitor cell isolation from umbilical cord blood based on immunoaffinity aqueous two-phase partitioning. *Biotechnol. Lett.*, **33** (12), 2373–7.

Sun, L., Zborowski, M., Moore, L.R. and Chalmers, J.J. (1998) Continuous, flow-through immunomagnetic cell sorting in a quadrupole field. *Cytometry*, **33** (4), 469–75.

Tan, S.J., Yobas, L., Lee, G.Y. *et al.* (2009) Microdevice for the isolation and enumeration of cancer cells from blood. *Biomed. Microdevices*, **11** (4), 883–92.

Tan, S.J., Lakshmi, R.L., Chen, P. *et al.* (2010) Versatile label free biochip for the detection of circulating tumor cells from peripheral blood in cancer patients. *Biosens. Bioelectron.*, **26** (4), 1701–5.

Tran, C.A., Torres-Coronado, M., Gardner, A. *et al.* (2012) Optimized processing of growth factor mobilized peripheral blood CD34+ products by counterflow centrifugal elutriation. *Stem Cells Transl. Med.*, **1** (5), 422–9.

Tsukamoto, M., Taira, S., Yamamura, S. *et al.* (2009) Cell separation by an aqueous two-phase system in a microfluidic device. *Analyst*, **134** (10), 1994–8.

Vahey, M.D. and Voldman, J. (2008) An equilibrium method for continuous-flow cell sorting using dielectrophoresis. *Anal. Chem.*, **80** (9), 3135–43.

Valero, A., Braschler, T., Demierre, N. and Renaud, P. (2010) A miniaturized continuous dielectrophoretic cell sorter and its applications. *Biomicrofluidics*, **4** (2), 022807.

VanDelinder, V. and Groisman, A. (2006) Separation of plasma from whole human blood in a continuous cross-flow in a molded microfluidic device. *Anal. Chem.*, **78** (11), 3765–71.

VanDelinder, V. and Groisman, A. (2007) Perfusion in microfluidic cross-flow: separation of white blood cells from whole blood and exchange of medium in a continuous flow. *Anal. Chem.*, **79** (5), 2023–30.

Velugotla, S., Pells, S., Mjoseng, H.K. *et al.* (2012) Dielectrophoresis based discrimination of human embryonic stem cells from differentiating derivatives. *Biomicrofluidics*, **6** (4), 44113.

Wang, K., Cometti, B. and Pappas, D. (2007) Isolation and counting of multiple cell types using an affinity separation device. *Anal. Chim. Acta*, **601** (1), 1–9.

Willoughby, N. (2009) Too big to bind? Will the purification of large and complex therapeutic targets spell the beginning of the end for column chromatography? *J. Chem. Technol. Biotechnol.*, **84** (2), 145–50.

Yamada, M. and Seki, M. (2005) Hydrodynamic filtration for on-chip particle concentration and classification utilizing microfluidics. *Lab Chip*, **5** (11), 123–9.

Yamada, M., Kano, K., Tsuda, Y. *et al.* (2007) Microfluidic devices for size-dependent separation of liver cells. *Biomed. Microdevices*, **9** (5), 637–45.

Yang, A.H. and Soh, H.T. (2012) Acoustophoretic sorting of viable mammalian cells in a microfluidic device. *Anal. Chem.*, **84** (24), 10756–62.

Zandstra, P.W., Bauwens, C., Yin, T. *et al.* (2003) Scalable production of embryonic stem cell-derived cardiomyocytes. *Tissue Eng.*, **9** (4), 767–78.

Zborowski, M., Fuh, C.B., Green, R. *et al.* (1995) Analytical magnetapheresis of ferritin-labeled lymphocytes. *Anal. Chem.*, **67** (20), 3702–12.

Zhang, B., Green, J.V., Murthy, S.K. and Radisic, M. 2012a. Label-free enrichment of functional cardiomyocytes using microfluidic deterministic lateral flow displacement. *PLoS One*, **7**(5), e37619.

Zhang, W., Kai, K., Choi, D.S. *et al.* (2012b) Microfluidics separation reveals the stem-cell-like deformability of tumor-initiating cells. *Proc. Natl. Acad. Sci. USA*, **109** (46), 18707–12.

Zheng, S., Lin, H., Liu, J.Q. *et al.* (2007) Membrane microfilter device for selective capture, electrolysis and genomic analysis of human circulating tumor cells. *J. Chromatogr. A*, **1162** (2), 154–61.

Zheng, S., Lin, H.K., Lu, B. *et al.* (2011) 3D microfilter device for viable circulating tumor cell (CTC) enrichment from blood. *Biomed. Microdevices*, **13** (1), 203–13.

6

Fundamental Points to Consider in the Cryopreservation and Shipment of Cells for Human Application

Glyn N. Stacey[1], Lyn Healy[2], Jennifer Man[1], Charles J. Hunt[1] and John Morris[3]

[1] *UK Stem Cell Bank, Advanced Therapies Division, National Institute for Biological Standards and Control (a centre of the Medicines and Healthcare products Regulatory Agency), Hertfordshire, UK*
[2] *Haematopoietic Stem Cell Laboratory, The Francis Crick Institute, London, UK*
[3] *Asymptote Ltd, St John's Innovation Centre, Cambridge, UK*

6.1 Introduction

Cryopreservation is the ability to hold cells at an ultralow temperature such that, at some point in the future, they can be returned to normal culture conditions and resume cell function, growth and division. It is a crucial technology for the life sciences and many biomedical therapies (Stacey and Day, 2014), but is a complex process and some of the mechanisms by which cells survive ultralow temperature preservation are yet to be fully understood. An intuitive challenge for cryopreservation is the formation of damaging intracellular ice, but the mechanisms of cryo-injury (Fowler and Toner, 2005) and cryoprotection are complex and it has been proposed that in some circumstances intracellular ice may have a cryoprotective effect (Acker and McGann, 2003).

Having preserved the cells successfully, their sustained viability is critical for reliable long-term storage and transport, where the cells may pass to different containers handled by different agents and authorities. Having appropriate procedures to assure successful preservation and sustained viability requires careful consideration of the full cold chain from cell source through to the final cell therapy products. Preservation of critical preparations, such as those intended for

Bioprocessing for Cell Based Therapies, First Edition. Edited by Che J. Connon.

transplantation, must be qualified to demonstrate that key features of the original cell preparation are retained on recovery. Careful implementation of the scientific principles of cryobiology is key to providing suitable protocols for use in industry and biomedicine. In this chapter we address key points that may need to be considered in the development of successful preservation methodologies and delivery of cell therapies.

6.2 The Role of Cryoprotective Agents (CPA)

A key feature of cryopreservation is the role of water and the need to adequately remove it from cells or eliminate its ability to form ice. Typically, when cells are frozen in aqueous suspension they are destroyed due to the formation of ice crystals. It was the discovery in the 1940s of the cryoprotective properties of glycerol (Polge et al., 1949), which led to the identification of a number of chemicals generically called cryoprotective agents (CPAs) capable of preserving cell viability at ultralow temperatures. It is important to recognise that the mechanisms involved in the action of CPAs are complex and are not fully understood. According to the commonly accepted theory of "colligative" action (Pegg, 2007), a key function of these CPAs in cell preservation is their ability to increase solute concentration both within the cell and extra-cellular, which suppresses intracellular ice formation. In order to do this, the so-called penetrating CPAs (e.g. DMSO, propanediol, methanol) must be able to cross the cell membrane readily and penetrate the cell without significant toxicity. This activity is thought to be significantly independent of the chemical characteristics of the CPAs. However, there is also a group of "non penetrating" CPAs (e.g. sucrose, trehalose), whose mode of action is thought to be at least in part due to their stabilising interaction with cellular membranes, a property also thought to impart the cryoprotective activity of certain large molecular weight compounds such as hydroxyethyl starch and polyvinyl propylene. Theoretical models of cryoprotection typically evoke the colligative theory, but it is clear that the full explanation of CPA action has yet to be established (for reviews, see Mazur, 2004 and Pegg, 2007).

Cryoprotectant molecules have biological activities beyond their cryoprotective properties. Some such as DMSO have a very wide range of effects on cells, including effects on membrane, cytoskeleton and gene expression, and may be toxic to cells on prolonged exposure (Notman et al., 2006). Also in vitrification methods cells may be exposed to very high CPA concentrations, which can also have significant toxic effects. During the development of new preservation protocols, it is therefore wise to carry out a cryoprotectant toxicity assay in which the cells are exposed to the CPA for varying times to see if there is any significant loss of viability or alteration of functionality.

6.3 Vitrification versus Cryopreservation

Cryopreservation is generally used to describe the technique of inducing and propagating extracellular ice that progressively dehydrates the cells, which then survive preserved in desiccated form within the ice. An alternative form of cell preservation, commonly called "vitrification", whereby the cell suspension is loaded with high levels of penetrating CPAs (often several in combination), induces a "glass" state in which the cellular and extracellular water cannot readily form ice crystals. When cell suspensions prepared in this way are then cooled very rapidly (1,000s of °C/min), the extreme viscosity results in a "glass" structure which prevents osmosis and the water molecules are unable to engage in the formation and propagation of ice. This technique has been widely used for complex structures including a variety of human plant and animal tissues, and provides a potential solution to preservation of any cell preparations where there is great variation in cell permeability for which standard cryopreservation would not be able to deliver the range of conditions required to optimally preserve viability in each of the component cell types.

6.4 Points to Consider in the Development of Cryopreservation Protocols

6.4.1 General Considerations

The development of a cryopreservation protocol that is robust and reliable requires careful evaluation of the cell system to understand what approach (e.g. preparation of cells, cryopreservation versus vitrification, type of cryoprotectant) would be likely to achieve the best recoveries of the therapeutic cells. It is important to recognise that a protocol which enables recovery of one cell type may lead to death of another and cryopreservation has even been used to eliminate undesirable cell types in cell preparations. Once a potentially useful combination of CPA and preservation protocol have been selected, it is important to optimise the individual steps of the procedure and understand how robust these are for the respective cell preparation. Here we will outline an approach for selection of suitable cryoprotocols and the key issues to be understood in order to deliver cells reliably for therapy.

6.4.2 Cellular Characteristics and Selection of Appropriate CPAs and Cooling Protocols

Cell size, permeability and morphology will have a significant influence on the method required for successful preservation and cells with complex structure, such as mature neurones and neutrophils, which may be difficult to cryoprotect by standard methods. Cell permeability is obviously an important factor which may vary depending on the cell type and may be increased when subjected to osmotic stress (Verkman et al., 1996). Cell preparations with diverse morphologies may also be difficult to cryopreserve successfully without the loss of cells for which the preservation method is less well suited, that is, the rate of osmosis of solutes may be significantly different due to inconsistency in cell size and/or the ratio of cytoplasm volume:cell surface area, that could result in inadequate desiccation and cryoprotection of some sub-populations. Where this may result in a significant loss of crucial cell types, careful characterisation of recovered cultures may be needed to assure sustained properties and sub-populations. If important populations are lost during cryopreservation, a vitrification method may be used which is not dependent on rates of osmosis and can therefore achieve equal preservation of a broad range of cell types.

It is also important to consider the possible impact of extreme desiccation on cells which have in principle a minimum volume of intracellular material, below which the cell cannot shrink any further by osmosis. As osmotic responses reduce cell volume, cell membranes become increasingly dehydrated which affects the nuclear and mitochondrial membranes as well as the plasma membrane. In addition, as membranes become increasingly folded to accommodate reduced volume, membrane fusion may occur which is also enhanced by some cryoprotectants (Notman et al., 2006). Furthermore, the cell may invaginate membrane as vesicles, which may cause severe osmotic stress and even cell lysis when the cell is thawed and rehydrated (Zade-Oppen, 1968).

Complications may also arise when preserving cells which remain attached to a culture surface or as a complex 3D structure (e.g. large aggregates, spheroids, organoids, tissue structures with extracellular matrix, scaffolds with cells attached) rather than in single cell suspension. Cell–cell contact in monolayer cultures is known to enable propagation of damaging ice through cell monolayers via gap junctions (Irimia and Karlsson, 2002) or alternative mechanisms (Higgins and Karlsson, 2013), including surface catalysed nucleation on the adjacent membrane surface to impinging ice (Acker et al., 2002). In addition, other effects may be observed due to the influence of extracellular matrix. Clearly, cells within tissues can behave differently from similar cells in a suspension with respect to intracellular freezing, and tissues or complex cell culture systems containing cells of different characteristics may need to be preserved using techniques which avoid ice formation such as vitrification (Pegg, 2010).

Animal cells may die due to a number of mechanisms for programmed cell death such as necrosis, apoptosis, extreme macroautophagy (Feng et al., 2014), anoikis (Frisch and Screaton, 2001) and other mechanisms including caspase 9 independent programmed cell death (Kroemer and Martin, 2005), depending on the nature of the insult to which they are subjected. During cryopreservation, a number of different effects can lead to such mechanisms of cell death. The most obvious of these is ice-crystal formation breaching the cell cytoplasmic membrane or other organelle membranes. Another key cause of loss of viability arises from the toxic effects of an increasing concentration of solutes in the residual water around the cells as ice forms during cryopreservation. Successful preservation is believed to be dependent on striking the right balance between negative effects of solute concentration and the positive effect of achieving a rate of cooling which permits osmosis to reduce intracellular water, thus decreasing susceptibility to ice formation. Most of the mechanisms described so far are associated with cryopreservation approaches. However, irrespective of whether cryopreservation or vitrification techniques are used, loss of viability may result from stress on the cell membrane during thawing, as osmosis may cause excessive swelling of the cells. Accordingly, care should be taken in recovery of cells to ensure viability is reproducible and does not subject cells to undue stress such as sheer or osmotic excursions. Furthermore, cells undergoing programmed cell death will interact with other cells and may also influence inflammation (by necrotic cells) or induce immunological anergy (by apoptotic cells) (Hotchkiss et al., 2009).

6.4.3 Key Events in Cryopreservation

During initial cooling, ice nucleation occurs in the significantly larger extracellular space, ice propagation causes it to become more concentrated and the cells begin to lose water by osmosis (Figure 6.1). The permeability of the cell membrane is critical to this process and is known to vary with temperature. Thus, it is important to set the rate of change of temperature so that the cells have time to allow osmosis to rebalance the changing tonicity of the environment and become progressively more dehydrated as the extracellular space becomes occupied with ice. Ultimately, the cells are held within a matrix of ice crystals, but are immediately surrounded by a residual aqueous phase with a high solute concentration. However, the preservation process is not complete until a sufficiently low temperature is reached to cause the residual solutes around the cells to become vitrified. This temperature is called the glass-transition temperature (Tg). The Tg is the temperature at which an amorphous material makes the reversible transition from a highly viscous state to a solid glass state.

Key factors to be considered in order to achieve reliable cryopreservation are:

1) ice nucleation (critical to initiation of removal of water from cells);
2) CPA concentration (to ensure suppression of intracellular ice formation, but which may also have significant biological effects, including cytotoxicity at high concentrations);
3) rate of cooling and cooling profile (to ensure balanced rate of dehydration of cells and to control the impact of the exothermic reaction of ice formation);
4) ratio of cells/total volume (high cell concentration can result in loss of viability, as described above);
5) storage conditions (to maintain cells below a critical temperature (Tg) in order to ensure a stable cryopreserved state); and
6) thawing procedure (to sustain viability of preserved cells and avoid damaging osmotic stress).

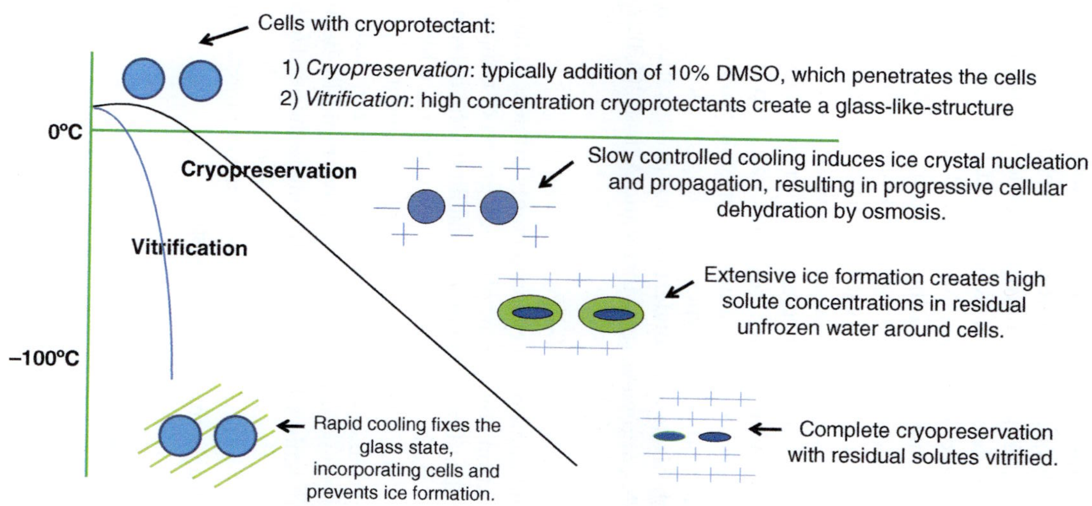

Figure 6.1 Changes in cells during cryopreservation and vitrification processes.

6.4.3.1 Ice Nucleation

As the temperature decreases in a cell suspension with an appropriate CPA, ice crystals begin to form, initiated by a process called nucleation. There are a numerous factors which may influence nucleation and this is probably one of the more difficult elements in a cryopreservation process to predict and control, and theories on ice-nucleation kinetics are still in development (Karlsson, 2010). In some protocols, nucleation is initiated by tapping the pre-cooled container of cryoprotected cells or adding a nucleating agent with the cryoprotectant. Inappropriate ice nucleation is a significant variable in controlled rate freezing methods for cryopreservation. It has the potential to result in variation of recoverable cell viability between samples and is thus potentially a major cause of inconsistency in banks of cells and cell-based medicines. Whilst much research has been carried out on nucleation in preservation of gametes and embryos, it has received less attention for other cell types including *in vitro* cell cultures. Further research needs to be undertaken for preservation of complex systems (for a recent review, see Morris and Acton, 2013).

6.4.3.2 CPAs: Concentration and Composition

In order to understand how cryoprotectants influence changes in the state of preservation conditions, phase diagrams of temperature versus CPA concentration can be used to facilitate the determination of key parameters, such as glass transition temperature, which is of value for establishing effective approaches to development of preservation and cryo-storage methods (for reviews, see Mazur, 2004 and Pegg, 2007). Combinations of multiple CPAs have typically been used for vitrification protocols, which can help to reduce the need for extremely high and toxic levels of a single CPA. They also enable the employment of both penetrating and non-penetrating cryoprotectants, which may have advantages in certain cell systems.

6.4.3.3 Cooling Rate

If the cooling rate is too fast, the cell membrane will not be able to allow osmosis fast enough and formation of potentially damaging intracellular ice is more likely to occur. If, however, the cooling rate is too slow, this will allow a cell osmotic response but will leave the cell exposed for longer to the increasingly toxic solutes before the sample becomes frozen and "cell viability" is lost. These two opposing lethal

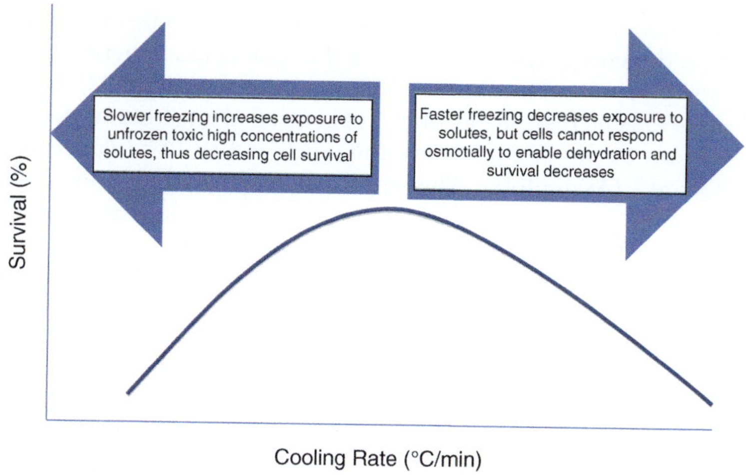

Figure 6.2 Generic "bell curve" illustrating typical trend in cell survival with increasing cooling rates.

effects have been called "Mazur's two phase hypothesis" (see references above), which is predicted from the bell-shaped profiles of graphs of cooling rate versus viability on recovery (Figure 6.2). Such graphs are useful to help identify an optimal cooling rate for a particular cell preparation (for examples, see Leibo, 1977 and Pegg, 2007). Typically, for animal cells this optimal cooling rate for DMSO cryoprotected cells falls between −0.5 and −5°C/min. Whilst a linear cooling rate in the preservation environment will enable cells to be cryopreserved effectively, the actual temperature within the cell suspension will not be linear, since at the point when ice forms there is a release of chemical energy known as "heat of crystallisation" (entropy of crystallisation). As this occurs, the temperature of the freezing suspension will rise and when most ice has formed the sample returns to its original cooling rate. Clearly, linear cooling rates are not necessarily ideal to assure sustained conditions for optimal osmotic dehydration of cells during cryopreservation. There are various events which may impact on a linear rate cooling environment, the most notable being the release of specific heat of crystallisation which typically occurs at around −12°C in cell preparations with cryoprotectant. This can be observed as a short-lived but significant increase in temperature in vials of cells as they are subjected to a −1°C/min linear cooling environment. A programmed increase in the cooling rate to compensate for this energy release can return the cooling rate experienced by the cell preparation to a linear −1°C/min (Massie et al., 2014; Morris et al., 1999), and there are a number of controlled rate freezing devices which can be pre-set to achieve this, such as Kry-10™ (Planer) and ViaFreeze™ (Asymptote). This approach may be needed for more demanding cultures or applications and will require mechanised cooling machines which can regulate the cooling rate.

It is important to ensure that a sufficiently low terminal temperature below the Tg is achieved to ensure that all residual unfrozen water is vitrified and therefore in a state suitable to sustain cell viability in long-term storage. Transfer from the cooling device to final destination in an ultralow temperature storage vessel should also be carefully controlled to ensure that vials are not significantly warmed to allow devitrification, which can happen remarkably quickly without any visible evidence of thawing. Typically the transfer temperature should be below −70°C for cryopreserved cells and below the glass transition temperature for vitrified material.

6.4.3.4 Storage of Cryopreserved Cells

It is important to sustain a sufficiently low temperature during storage to keep the cell suspension in a stable frozen state below Tg (see above) and it is generally accepted that such storage will assure stable maintenance of viable preserved cells for several decades at least (Day and Stacey, 2008). Ultralow temperature (−150°C) mechanical deep freezers are commercially available and are successfully used by some cell therapy companies. However, it is generally considered that the most stable conditions for storing cells at ultra-low temperatures are provided under liquid nitrogen. Storage above liquid nitrogen in its gaseous phase (so-called "vapour-phase" storage), is often recommended to reduce the risk of viral cross-contamination (Fountain et al., 1997; Grout and Morris, 2009), which has led to deaths in bone marrow transplant patients (Tedder et al., 1995). However, such storage increases the likelihood for temperature instability from the inherent temperature gradient between the surface of the nitrogen liquid and the upper regions of the storage vessel. Failure to maintain liquid nitrogen levels in storage tanks or removal of vial racks temporarily from storage vessels can cause significant and potentially damaging warming above the Tg. If this occurs, cells may be exposed to unfrozen high concentration solutes. In the case of cells cryopreserved by controlled rate freezing, the Tg is usually achieved once −125°C is reached. However, for vitrified material, mobilisation of water, and thus ice formation, may occur at much lower temperatures of around −130°C. Cryopreserved cells may survive for short periods at higher temperatures such as −80°C for the purposes of quarantine or shipment, but this cannot be used for long-term storage (Massie et al., 2013). For vitrified material, temperatures above the glass transition temperature, even if transient, should be avoided to prevent formation of ice crystal nuclei that may grow on repeated warming under such conditions. Accordingly, it is important to assure that stored vials are stable and that the temperature of the gaseous phase in storage vessels is maintained above the Tg, by regular or automatic filling of vessels and that they are constantly monitored. Ideally, vessels should be fitted with automatic filling devices. In addition, any temporary removal of vials from storage into ambient temperatures should be strictly minimised.

6.4.3.5 Thawing and Recovery of Frozen Cells

The process of thawing and recovery is as important as the cooling element of preservation. As with cooling, successful thawing must be carried out at a rate that allows cell membranes to permit osmotic rebalancing, as extracellular ice becomes aqueous water and the intracellular environment is still highly concentrated. This process will cause cells to take up water and swell, thus putting stress on cells which are metabolically and structurally compromised. This state means that the cells are initially non-functional and without the usual cytoskeleton organisation especially in cells harvested enzymatically. This effect will be especially marked if the extracellular CPA is diluted too quickly by rapid addition of growth medium. Cells in these conditions should not be exposed to unnecessary sheer forces such as vigorous shaking or pipetting and addition of culture medium step wise may help to reduce cell damage.

Having recovered the cells, it is also important to remember that the preservation process may have caused loss or alteration of certain populations within the original cell sample. Testing must therefore be carried out to demonstrate that the recovered culture is equivalent to the original. Tests for viability are important and should be suitable for the cell or culture type (see below) and should not be relied on to imply any general functional competency. Supplementary characterisation should be applied to address cell functionality (USP, 2014). It is important to note that results of viability tests carried out immediately post-thaw may overestimate the actual recovery and viability of cells, for example the retention of membrane integrity whereby a trypan blue dye exclusion assay will indicate the cell is viable, even when it has irrevocably committed to apoptosis and cell death. Thus, viability measurements should be carried out not just on thawing, but also some time after thawing, for example 24–48 h for cell lines.

A wide range of techniques have been used to measure cell viability and these assess a very wide range of different cellular attributes. The parameters include membrane enzymes (fluorescein diacetate), membrane lipid degradation (malondialdehyde), dye exclusion by the plasma membrane (Trypan blue, eosin), organelle function (Neutral Red), tetrazolium reduction (MTT), resazurin reduction (Alamar blue), and expression of apoptosis associated genes. Each assay will give specific perspective on cellular biochemistry, physiology or function and will have its own individual strengths and weaknesses for different cell types and circumstances. The selection of a viability assay should therefore be made on the basis of the cell culture characteristics, its functional potential and the specific application of the cells in question, on a case by case basis.

6.5 Large-volume Freezing

The nature of ice formation and propagation during cryopreservation can be crucial to the post-thaw survival and performance of a preparation of cells. The water–ice phase transition can be dominated by one of two processes; network solidification or progressive solidification. Network solidification is typically present in small sample vials or cryo-straws, whereas progressive solidification is more often observed in larger volumes. Scale-up from small volume to large volume preservation formats for cryo-banking protocols will necessitate detailed investigation when considering critical applications such as cell therapy where safety and process consistency are critical (Massie et al., 2014). Progressive solidification should not prove to be a significant hurdle in the development of successful protocols for cryopreservation in large volumes, although total viable cell recoveries and functional assessments of preserved cells may differ between samples cooled with either progressive or network solidification (Kilbride et al., 2014).

6.6 Cryopreservation as Part of Manufacturing Processes

6.6.1 Containers for Cryopreserved Cells

For cell suspensions, the choice of primary container will generally be conditional on the mode of cryopreservation. For cell preparations, including spheroids, aggregates and encapsulated cells, the most practical and generally acceptable options for biomedical applications include straws, vials and bags. Each option should be assessed for its suitability, not only for the mode of cryopreservation (e.g. are the cooling rates required achievable, is the mode of ice development suitable (see Section 6.5 above)), but also its ability to reduce the risk of contamination (primarily during cooling and storage) and to be provided in sterile form. An additional requirement is compliance with regulatory guidelines, including absence of bioactive leachates from vial components and requirements for labelling of the primary container (Andrews et al., 2015). The use of open systems (such as open straws) or cryovials with external vial screw threads, represent a risk of contamination to stored cells and is not consistent with best practice.

6.6.2 Controlled Rate Freezers and Storage Systems

Cryopreservation protocols used to manufacture starting materials and products, as for all other crucial elements of the manufacturing process, must comply with Good Manufacturing Practice (GMP) and the cooling devices used in preservation must also be validated for performance under the manufacturing licence. Controlled rate freezers (CRFs), which achieve cooling by injection of liquid nitrogen into a chamber containing the cell preparations, generate high levels of particles in the environment which would invalidate the strict requirement for very low particle levels in the clean room environment (ISO, 1999). If such devices are used outside the production area, the means by which cells are moved to the CRF (to ensure that CPA exposure time/temperature does not compromise cell survival or lead to contamination) and subsequently from the CRF to a qualified storage system, need to be established and validated.

Liquid nitrogen-free CRFs, such as those employing the Stirling Cycle principle, may provide an alternative (Morris et al., 2006, 2010). Such equipment may provide clean room-compatible options for controlled rate freezing; nevertheless, they will need to be assessed for their ability to achieve the required scale (unit volume and number) and cooling rates appropriate to the cell preparations to be preserved.

It is not currently possible to store cells at sufficiently low temperatures without the use of liquid nitrogen, the boil-off from which creates high levels of particulates in the laboratory air, which are incompatible with operation in a clean room area where cells for use in human therapy are cultured (maintained at air particle levels A and B) or even annexes to these areas (typically Grade D and above) (ISO14644, 1999). Thus, when handling cells for clinical application, strategies are needed to enable storage in liquid nitrogen in areas with "unclassified" air quality, but which maintain asepsis of the preserved cells. Potential solutions include heat sealable vial "sleeves", sealed storage containers, etc.

Sample labelling and traceability is crucial to effective and safe management of biobanks for clinical use. The American Association of Tissue Banks has implemented the Information Standard for Blood and Transplant (ISBT) 128 system (www.iccbba.org) and this is now also being adopted in other countries.

6.6.3 The Cold-chain: Challenges and Solutions

Assuring that cryopreserved cells are maintained in a consistent state from preservation to delivery at the clinical site, is clearly crucial for efficacy and safety for cell–base medicines. Temperature logging of materials during storage, shipment and delivery will also be a key requirement to demonstrate that no change could have occurred in the cells. Cells are at particular risk during transport and thus all aspects of shipment must be controlled and qualified including validation of shipment containers, use of temperature monitors, implementation of service level agreements with shipping companies and formal arrangements to deal with adverse events during shipment, such as delays which will require emergency arrangements to recharge liquid nitrogen shippers. Conditions to which cells will be exposed during shipment should be carefully evaluated. Airfreight shipment will be a common means of shipment of cell-based products and brings some challenges, including the higher levels of radiation experienced due to airport screening using X-rays and the elevated levels of radiation experienced at high altitude. However, risk analyses show that the likely levels of irradiation experienced in air travel are unlikely to present levels of exposure of concern compared to the doses required to cause disease in humans (personal communication Dr R Khan, PHE, UK).

6.6.4 Biobanking and Regulatory Requirements

It is clear that professional biobanking systems must address a number of issues related to implementation of cryopreserved storage (Benson et al., 2013). Guidance on biobank management has been published (ISBER, 2012), as has international consensus on the biobanking starting materials for cell therapies (Andrews et al., 2015). For tissues and cells intended for clinical use, there are also more specific regulatory demands. In Europe there is specific legislation for the import and export of tissues (such as the Human Tissues Authority code of practice for import of cells for human application, www.hta.gov.uk). The overall requirements for preparation of starting materials, procurement, processing, testing and storage of cells for human application are outlined in the technical annexes of the European Union Tissues and Cells Directive. However, the situation is highly variable around the world. In some countries, a simple statement of commercial worth is required, whereas in Taiwan for example there are specific import and export regulations, and in other countries, these issues are still under consideration (see tabulated regulatory information in Andrews et al., 2015). Cells cryopreserved by slow cooling may be transported in "dry-ice" (solid carbon dioxide at −79°C), provided that they are not kept at this temperature for long periods. Vitrified material should not be transported in dry ice as, for the reasons explained in previous sections, devitrification and cell damage will result. Cells cryopreserved by either method may be transported in liquid nitrogen charged "dry-shippers", which are probably the most secure form of ultra-low temperature vessel for transport.

Further expert opinion on preservation of cells for clinical application can be found in the international consensus on stem cell banking (Andrews et al., 2015) and the USP informational general chapter "Cryopreservation of Cells" (USP, 2014).

6.7 Conclusions

Traditionally, cryopreservation has been applied in a largely empirical way, applying methods which have worked for similar cells or tissues, without evaluating the actual needs of each new cell type or preparation or gaining an understanding of how the preservation process works or how efficient it is in the new cell type. Significant loss of viability (in the order of log reductions in the case of virus preservation, unpublished data on the preservation of virus controls for in vitro cell culture assays, G Stacey) has often been tolerated for microorganisms and cultures used for research, provided the cells recover and grow adequately for experimental work. Now, in more critical applications such as cell therapy, where the biological systems are very complex and varied, there is a significant unmet need for preservation protocols which can assure more reliable and reproducible recovery of cell cultures. In many instances, such therapeutic cell preparations will need to retain the functionality of multiple subpopulations, which may be diverse in their morphology and permeability and may require new approaches to preservation other than commonly used controlled rate freezing methods, including modified vitrification techniques and use of novel antifreeze proteins.

Acknowledgements

The development of this manuscript was an output of a project entitled "Development of a GMP cryogenic cold chain for clinical delivery of regenerative medicine therapeutics", funded under the Technology Strategy Board Regenerative Medicine Programme – Tools and Technologies (Project no 101103) and the Phase IV UK Stem Cell Bank project ref. MR/L01324X/1.

References

Acker, J.P. and McGann, L.E. (2001) Membrane damage occurs during the formation of intracellular ice. *Cryo-Letters*, **22**, 241–54.

Acker, J.P. and McGann, L.E. (2002) Innocuous intracellular ice improves survival of frozen cells. *Cell transplantation*, **11** (6), 563–71.

Acker, J.P. and McGann, L.E. (2003) Protective effect of intracellular ice during freezing? *Cryobiology*, **46**, 197–202.

Andrews, P.W., Baker, D. et al., on behalf of the International Stem Cell Banking Initiative 2015. Points to consider in the development of seed stocks of pluripotent stem cells for clinical applications: International Stem Cell Banking Initiative (ISCBI). *Regenerative Medicine*, **10**(2)s, doi: 10.2217/RME.14.93

Benson, E.E., Betsou, F., Fuller, B.J. *et al.* (2013) Translating cryobiology principles into trans-disciplinary storage guidelines for biorepositories and biobanks: A concept paper. *CryoLetter*, **34**, 277–312.

Day, J.G. and Stacey, G.N. (2008) Biobanking (invited review). *Mol. Biotechnol.*, **40**, 202–13.

Day, J.G. and Stacey, G.N. (2007) in *Cryopreservation and Freeze-drying Protocols* (eds D.G. Day and G. Stacey), Humana Press, Totowa, US.

Feng, Y., He, D., Yao, Z. and Klionsky, D.J. 2014. The machinery of macroautophagy. *Cell Res.*, **24**, 24–41. Epub 2013 Dec 24.

Fowler, A. and Toner, M. (2005) Cryo-injury and biopreservation. *Ann. N.Y. Acad. Sci.*, **1066**, 119–35.

Frisch, S.M. and Screaton, R.A. (2001) Anoikis mechanisms. *Current Opinion in Cell Biology*, **13**, 555–62.

Fountain, D., Ralston, M., Higgins, N. *et al.* (1997) Liquid nitrogen freezers: A potential source of microbial contamination of hematopoietic stem cell components. *Transfusion*, **37**, 585–91.

Grout, B.W.W. and Morris, B.J. (2009) Contaminated liquid nitrogen vapour as a risk factor in pathogen transfer. *Theirogenology*, **71**, 1079–82.

Hotchkiss, R.S., Jonathan, A., McDunn, E. and Swanson, P.E. (2009) Cell death in disease: Mechanisms and emerging therapeutic concepts *N. Engl. J. Med.*, **361**, 1570–83.

Higgins, A.Z. and Karlsson, J.O. (2013) Effects of intercellular junction protein expression on intracellular ice formation in mouse insulinoma cells. *Biophys. J.*, **105**, 2006–15.

Irimia, D. and Karlsson, J.O.M. (2002) Kinetics and mechanisms of intracellular ice propagation in a micropatterned tissue construct. *Biophys. J.*, **82**, 1858–68.

International Organization for Standardization 1999. ISO 14644. Clean rooms and associated controlled environments. Part 1: Classification of air cleanliness. International Organization for Standardization. (NB for other more recent parts to this guidance, see http://www.iso.org/iso/iso_catalogue/catalogue_tc/catalogue_tc_browse.htm?commid=54874)

ISBER (2012) Best practices for repositories: Collection, storage, retrieval, and distribution of biological materials for research. *Biopreservation and Biobanking*, **10**, 79–161.

Karlsson, J.O. (2010) Effects of solution composition on the theoretical prediction of ice-nucleation kinetics and thermodynamics. *Cryobiology*, **60**, 43–51.

Kilbride, P., Morris, G.J., Milne, S. *et al.* (2014) A scale-down process for the development of large volume cryopreservation. *Cryobiology*, **69**, 367–75.

Kroemer, G. and Martin, S.J. (2005) Caspase-independent cell death. *Nature Medicine*, **11**, 725–30.

Leibo, S. (1977) Fundamental cryobiology of mouse ova and embryos, in *In: The Freezing of Mammalian Embryos. Ciba Foundation symposium 52 (new series)*, Elsevier, Amsterdam, The Netherlands, pp. 69–92.

Massie, I., Selden, C., Hodgson, H. and Fuller, B. 2013. Storage temperatures for cold-chain delivery in cell therapy: A study of alginate-encapsulated liver cell spheroids stored at –80°c or –170°c for up to 1 year. *Tissue Eng. Part C Methods*, **19**, 189–95. Epub 2012 Sep 7.

Massie, I., Selden, C., Hodgson, H. et al. 2014. GMP cryopreservation of large volumes of cells for regenerative medicine: active control of the freezing process. *Tissue Eng. Part C Method.*, **20**, 693–702. Epub 2014 Feb 24.

Mazur, P. (2004) Principles of cryobiology, in *In: Life in the Frozen State* (eds B. Fuller, N. Lane and E.E. Benson), CRC Press, Boca Raton, FL, London, New York, Washington DC, pp. 3–66.

Morris, G.J. and Acton, E. 2013. Controlled ice nucleation in cryopreservation – a review. *Cryobiology*, **66**, 85–92. Epub 2012 Dec 12.

Morris, G.J., Acton, E., Collins, D. *et al.* (2010) Optimisation of current Good Manufacturing Practice (cGMP) compliant controlled rate freezing for human embryonic stem cells. *Cryobiology*, **61**, 406–14.

Morris, G.J., Acton, E., Faszer, K. *et al.* (2006) Cryopreservation of murine embryos, human spermatozoa and embryonic stem cells using a liquid nitrogen-free, controlled rate freezer. *Reprod. Biomed. Online*, **13**, 421–6.

Morris, G.J., Acton, E. and Avery, S. (1999) A novel approach to sperm cryopreservation. *Hum. Reprod.*, **14**, 1013–21.

Notman, R., Noro, M., O'Malley, B. and Anwar, J. (2006) Molecular basis for dimethylsulfoxide (DMSO) action on lipid membranes. *J. Am. Chem. Soc.*, **128**, 13982–3.

Pegg, D.E. (2007) Principles of cryobiology, in *In: Cryopreservation and Freeze-drying Protocols*, 2nd edn (eds J.G. Day and G. Stacey), Humana Press Totowa, US, pp. 59–72.

Pegg, D.E. 2010. The relevance of ice crystal formation for the cryopreservation of tissues and organs. **60**, S36–44. Epub 2010 Feb 14.

Polge, C., Smith, A.U., Parkes, A.S. *et al.* (1949) Revival of spermatozoa after vitrification and dehydration at low temperatures. *Nature*, **164**, 666–76.

Stacey, G.N. and Day, J.G. (2014) Putting cells to sleep for future science. *Nature Biotechnology*, **32**, 320–2.

Tedder, R.S., Zuckerman, M.A., Goldstone, A.H. *et al.* (1995) Hepatitis B transmission from contaminated cryopreservation tank. *Lancet*, **346** (8968), 137–40.

USP 2014. Available at US Pharmacopeial Forum: www.usp.org/usp-nf/pharmacopeial-forum

Verkuna, A.A., van Hoek, A.N., Ma, T. *et al.* (1996) Water transport across mammalian cell membranes. *Am. J. Physiol.*, **270**, C12–30.

Zade-Oppen, A.M.M. (1968) Post-hypertonic heamolysis in sodium chloride systems. *Acta Physiol. Scand.*, **73**, 341–64.

7

Short-term Storage of Cells for Application in Cell-based Therapies

Stephen Swioklo and Che J. Connon

Institute for Genetic Medicine, Newcastle University, Newcastle upon Tyne, UK

7.1 Introduction

Cell-based therapies encompass a number of approaches that aim to regenerate damaged tissue, or to accelerate wound healing by exploiting the properties of certain cell types. Whether this is through the integration and repopulation of cells into a site of injury, or regeneration aided by the wealth of paracrine factors produced by cells, their potential for treating a vast number of diseases is considerable. Cell-based therapies can be divided into two major categories: i) autologous, using a patient's own cells, or ii) allogeneic, using donor cells. The interest in both of these categories is substantial, with a growing body of evidence suggesting their clinical efficacy, coupled with the emergence of approved cell therapy products (CTPs) to the market. Accompanying this, however, are unmet bioprocessing and logistical challenges associated with the storage and distribution of cells between sites of manufacture and the clinic. This chapter explores the logistical landscape involved in the manufacture and clinical delivery of CTPs, highlighting areas where short-term storage of cells could be of benefit. It examines hypothermic (0–32°C) storage of cells as a means to achieve this, discussing the optimal conditions for cell survival and function, and its potential application in the cell therapy industry.

Bioprocessing for Cell Based Therapies, First Edition. Edited by Che J. Connon.

7.1.1 Advances in Cell-based Therapies

The clinical application of cell-based therapeutics is ever accelerating, with almost 5,000 clinical trials being registered globally between 1992 and 2012, and the number of novel clinical trials demonstrating a dramatic increase in the last decade. These include both autologous and allogeneic cell therapies targeted towards a wide range of conditions, including cardiovascular disease, neurodegenerative diseases, cancer, liver disease, diabetes, skeletal disorders and eye diseases (Foley and Whitaker, 2012; Li et al., 2014). Of note, there is much attention focusing on mesenchymal stem cell (MSC)-based therapies, accounting for 63% of novel allogeneic therapies being developed (Li et al., 2014). Epitomising this growth is Prochymal® (Osiris), an allogeneic CTP that exploits the anti-inflammatory properties of MSCs. Prochymal® has effectively progressed through clinical trials, with on-going Phase III clinical trials for graft versus host disease and Crohn's disease, and Phase II clinical trials for acute myocardial infarction and type I diabetes. Furthermore, a number of other CTPs are in phase III of development, including treatments for limb ischaemia, chronic myocardial ischaemia, complex perianal fistula and congestive heart failure (recently reviewed in Syed and Evans, 2013). Accompanying the growing intensity of clinical research is the emergence of CTPs on the market, with a number being granted either Food and Drug Administration (FDA) – or European Medicines Agency (EMA) – approval including:

- ChondroCelect® (TiGenix) and Carticel® (Genzyme) for autologous knee cartilage repair;
- Epicel (Genzyme) with "Humanitarian Device" status as an epidermal autograft for treating burns;
- Provenge® (Dendreon), an autologous immunotherapy for prostate cancer treatment; and
- Dermagraft® and Apligraf® (Organogenesis) for dermal ulcer allogeneic grafts.

With the total stem cell therapy market of 8.8 billion US dollars predicted for 2016 (Syed and Evans, 2013), the development of CTPs is rapidly expanding.

7.1.2 The Logistical Landscape for CTPs and the Requirement for Short-term Storage of Cells

One of the major bottlenecks for CTPs is in the short-term storage of cells for quality, safety and efficacy testing, and their distribution from the donor or bank, to sites of processing, and back to the clinic (Pamphilon et al., 2010; Robinson et al., 2014). The complex logistical supply chain could involve three or more manufacturing sites (Hourd et al., 2014), making careful consideration of the effect of storage/distribution conditions on cell viability and therapeutic function paramount. The strategy through which this is carried out is likely to differ, depending on whether the CTP is allogeneic or autologous. Whilst allogeneic CTPs would require the mass cryopreservation of expanded cells and their storage in banks, patient-specific autologous cell production would involve processing enough cells for a single treatment, or course of treatments, potentially obviating the requirement for cryopreservation. This could be achieved through a centralised production system whereby autologous cells are processed at a good manufacturing practice (GMP) facility, or facilities, located away from the patient's primary care facility. Alternatively, there is an argument for a "GMP in a box" scenario, whereby patient-specific cells could be produced within the primary care facility (argued by Trainor et al., 2014). With the advent of in-house cell processing systems such as Cytori's Celution® for adipose-derived stem cell processing, this approach appears promising. Having individual cell processing systems within primary care facilities however, may be costly, and cells may still require a period of storage following production and quality assurance due to scheduling and clinician availability, as well as if the therapy involves a course of multiple treatments (Parkins et al., 2006).

Although cryopreservation is likely to be required for the long-term storage of cells, there are a number of biological and technical issues associated with its use. One of the major issues is the requirement for cryoprotectant agents (CPAs), most commonly dimethyl sulfoxide (DMSO), which at the concentrations used, can be toxic to cells, has been associated with a number of adverse effects in clinical studies, and can be costly and complicated to remove (recently reviewed in Thirumala et al., 2013). Freezing media also commonly contains foetal bovine serum (FBS), which could be a source of prion, viral and zoonotic diseases, and could be immunogenic due to xenogeneic serum antigens (reviewed in Mannello and Tonti, 2007).

It should be mentioned that much effort has been focused on reducing concentrations of DMSO in freezing media, as well as eliminating xenogeneic serum (Deb, 2009; Schulz et al., 2012; Seo et al., 2011; Zeisberger et al., 2011), but an existing concern remains. Cryopreservation has also been associated with the activation of apoptotic and necrotic pathways 6 to 24 hours following post-thaw culture, with resultant loss of cell viability and function (reviewed in Bissoyi et al., 2014). This introduces variability into the final CTP, unfavourable for GMP compliance. In addition, the long-term storage of cells in liquid nitrogen is expensive, requiring the constant monitoring of levels and supply as well as specialist equipment and facilities. There is also the risk of contamination with viruses or pathogens, and the production of liquid nitrogen itself has a high carbon footprint (reviewed in Loi et al., 2013). It is also important to consider the risk of thawing during shipping and, where the clinic may not have access to cryogenic freezers, "just-in-time" shipping must be employed where there is a limited time window for transplantation. The hold-time subsequent to thawing of cells could be limited to as little as 2 hours before experiencing a loss in viability, as has been demonstrated with MSCs (Pal et al., 2008). Therefore, the shipment of cryopreserved cells will inherently involve tight control of dispatch from centres of processing to the clinic where both clinicians and the patient will need to be equipped and prepared to receive the CTP. Due to the detrimental effects of cryopreservation and thawing, it has been argued that it is advantageous to maintain cells in their liquid state (Pamphilon et al., 2010) and, in a recent assessment by experts in the field, it was considered critical that improved preservation and transport enabling technologies be developed in order to eliminate GMP-regulated process steps of preservation removal and cell thawing (Hourd et al., 2014).

Liquid-state storage, whilst avoiding many of the issues associated with cryopreservation, still shares concerns over limited time windows where difficulty in timing the transfer of cells to the clinic due to a restricted, inflexible shelf life can result in a high level of wastage of materials (Hernon et al., 2006). Nevertheless, CTPs that have been granted regulatory approval generally adopt liquid-state storage, and are produced in relatively small unit numbers. Their short shelf life means that they are produced 'just in time' rather than being stockpiled. This keeps production costs relatively low but it does mean that the cost of goods made as a result of economies of scale is limited (Kemp, 2006). ChondroCelect® (TiGenix), an EMA-approved CTP, delivers autologous chondrocytes for knee cartilage repair. Four million cells are stored at 15–22°C in sealed vials containing 0.4 mL DMEM for up to 48 hours (ChondroCelect: EPAR-product information) and, similarly, Carticel® (Genzyme), at a dosage of 12 million cells, is shipped within secondary packaging capable of maintaining an appropriate temperature (not stated) for up to 72 hours, and should be held at room temperature before required (Carticel: Prescribing Information). Provenge® (Dendreon) delivers autologous activated antigen presenting cells and is shipped in an insulated polyurethane container containing gel packs to maintain an appropriate temperature (not stated), and should be held at room temperature for no longer than 3 hours prior to infusion (Provenge: Prescribing information) and Epicel® (Genzyme) epidermal keratinocyte autografts are placed in serum-free DMEM packed within temperature controlling packaging and can be stored for up to 24 hours in a "cool room temperature" (e.g. 13–23°C) (Epicel: Directions for use). Similarly, Apligraf® (Organogenesis), allogeneic dermal fibroblast and keratinocyte grafts, are stored at temperatures between 20 and 23°C in an agarose nutrient medium, but have a shelf life of 15 days (Apligraf: APG Factsheet).

With the exception of Dermagraft®, cryopreserved dermal fibroblast allografts packaged in a cryoprotectant solution containing DMSO and bovine serum (Dermagraft: Directions for use), the majority of CTPs utilise hypothermic storage, a mechanism whereby a living biologic is maintained in a suspended state at temperatures above 0°C but below the mammalian normothermic range of 32–37°C (Baust, 2006). The majority of shelf lives, however, are relatively limited, with only one CTP (Apligraf®) having a shelf life extending past 14 days, the period of time required for full United States Pharmacopeia (USP)-sterility testing (Naughton, 2002), introducing an element of risk. In fact, at least in the case of suspended cells for implantation, shelf lives appear restricted to 3 days at most, representing a relatively limited time window.

Extending the shelf life of hypothermally stored CTPs, whilst preserving viability and therapeutic potency, offers a number of benefits at multiple stages between processes for both autologous and allogeneic therapies (summarised in Figure 7.1). Autologous processes are shown in Box A, where (i) tissue is taken from the patient at their primary care facility. This may be a single sample or multiple samples requiring a period of hypothermic storage before (ii) sample processing that may occur on-site or may require hypothermic transport of the unprocessed tissue to a GMP facility. (iii) shows that the processed tissue may require hypothermic transport to a GMP facility for cell number up-scaling. Box B (i) shows that following up-scaling, the CTP must be purified, formulated and packed. Hypothermic storage may be required prior to dispatch to allow for full quality assurance testing. (iia) shows that autologous cells may be cryopreserved and banked prior to storage with potential application of hypothermic distribution between GMP facilities and

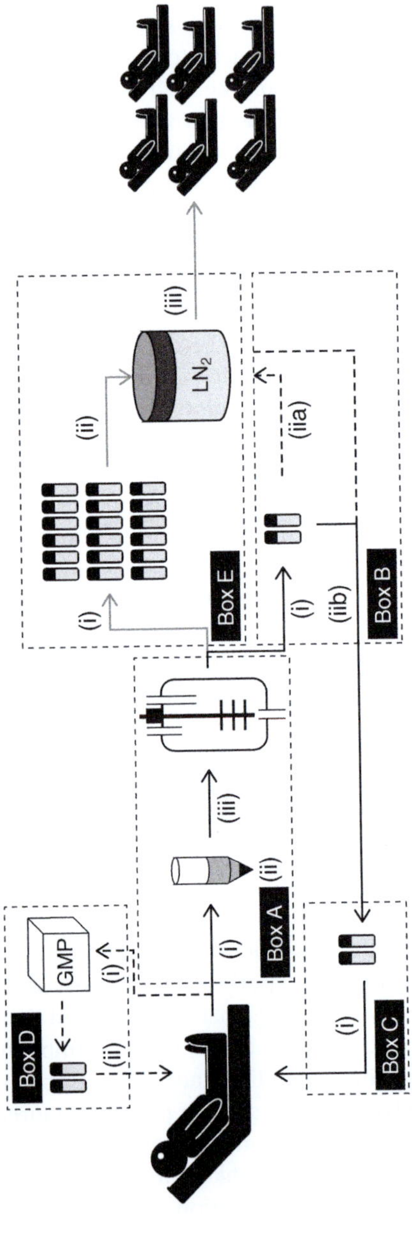

Figure 7.1 The potential for hypothermic storage in CTP logistics for autologous and allogeneic therapies.

back to the clinic. Autologous cells will more likely be maintained in their liquid state (iib) and require hypothermic transport from GMP facilities to the clinic. Box C (i) shows that patient/clinician scheduling and potential multiple dose regimes would require an extended period of hypothermic storage prior to patient delivery. Alternatively, Box D represents a "GMP in a box" scenario, whereby a single hermetic bioprocessing system could be situated within the primary care trust to (i) process patient tissue for (ii) patient delivery. Quality assurance testing, scheduling and multiple doses may necessitate a period of hypothermic storage.

The allogeneic processes shown in Box A, where tissue will be processed as per autologous CTPs, involves more extensive up-scaling, with similar requirements for hypothermic storage. Box E (i) shows that following up-scaling, large numbers of cells must be purified, formulated and packed. Extensive quality assurance testing will be required for allogeneic cells, potentially requiring hypothermic storage prior to (ii) cryopreservation and banking, which could be at a different GMP facility requiring hypothermic transport. (iii) shows a system by which ultra-cold storage and hypothermic storage are integrated to allow distribution from banks to the clinic with an appropriate time window to allow for quality assurance testing, scheduling and multiple doses.

Not only does it hypothermic storage have the potential to extend the period for quality assurance, sterility testing, characterisation and efficacy testing, it also reduces the risk of wastage due to inadequate continuity between sites of processing and patient scheduling at the clinic. In addition, where a course of therapy requires multiple tissue samples or dosing, tissue or cells, respectively, can be stored at the clinic obviating the requirement for multiple shipments. Because of this, much effort has been focused on optimising technologies for the storage and distribution of therapeutic cells in their liquid state at hypothermic temperatures that are simple, scalable and do not affect the viability or function of therapeutic cells.

7.2 Hypothermia and Mammalian Cell Storage

Based upon the principle that a reduction in temperature will decrease metabolic rate and, thus the demand for oxygen and nutrients, low temperatures have long been exploited in surgery where blood is restricted, with wide clinical use in cardiac surgery since the early 1950s (Swan et al., 1955). Subsequently, this was extended to the storage of organs and the development of early hypothermic biopreservation solutions in the late 1980s, namely Viaspan®, also known as University of Winconsin (UW) organ preservation solution. This was originally developed for organ storage at 0–5°C (e.g. refrigerated), formulated with a number of cell impermeant agents to prevent cell swelling, as well as glutathione and adenosine to augment antioxidant capacity and ATP production respectively upon reperfusion. Viaspan® was successfully used for the short-term storage of a number of whole organs including kidney, liver and pancreas (reviewed in Southard and Belzer, 1995) and later for isolated cells (Olinga et al., 1997).

Indeed, the principle that was first considered for surgery whereby reduced temperature could evoke a decrease in metabolic requirements is directly applicable to cells, where hypothermic conditions can evoke a state of "suspended animation"; a process recently termed "cell pausing" (Robinson et al., 2014). In a coordinated cellular response involving modulations in transcription, translation, metabolism, cell cycle progression and cytoskeletal organisation (reviewed in Al-Fageeh and Smales, 2006), cells can effectively respond to hypothermic temperatures to become suspended in time, a process of great potential application in the storage of cells for a number of applications. As a consequence of this, there has been considerable interest in the development of commercially available specialised hypothermic preservation media, predominantly for storage at 0–5°C, including HypoThermosol® from BioLife Solutions (Rhoads, 1994), Optisol-GS from Bausch and Lomb (Raeder et al., 2007), and Unisol™ from the Biopreservation Company (Campbell et al., 2013). Accompanying this is a growing body of literature describing novel methods for hypothermic preservation, as well as the limitations to such approaches. This chapter will discuss the optimal conditions for hypothermic cell preservation, with special reference to mesenchymal stem cells (MSCs), and the feasibility of its implementation in the storage and distribution of cell-based cell therapies.

7.2.1 Hypothermic Storage of Mesenchymal Stem Cells (MSCs)

MSCs have considerable potential in cell-based therapy, with a notable push for clinical application in recent years (Li et al., 2014). The reasons for this are multiple:

i) they are readily accessible, being present in a number of tissues, most commonly accessed from bone marrow, adipose tissue and umbilical cord;

ii) they have a tri-lineage mesodermal differentiation capacity, as well as having an ability to differentiate into non-mesodermal lineages;
iii) they produce a number of trophic, anti-inflammatory, immunomodulatory and anti-apoptotic factors in an environmentally-responsive manner; and
iv) they are considered immune-privileged due to low to medium expression of Human Leukocyte Antigen (HLA) class I and II molecules resulting in feasible application in allogeneic cell therapy (see Murphy et al., 2013 for a recent review).

For these reasons, MSCs have been investigated for a number of therapeutic applications, including ischaemia revascularisation, cardiovascular tissue regeneration, bone/cartilage repair, and wound healing (reviewed in Pikula et al., 2013). Common to other cell types however, there exist the logistical complications for their storage and distribution as presented in the introduction to this chapter. Subsequently, much research has focused on the hypothermic preservation of MSCs.

Perhaps the first studies to examine hypothermia in MSC preservation were by Muraki *et al.* (2006), where the suspension of bone marrow-derived human mesenchymal stem cells (BM-hMSCs) in PBS (1×10^6 cells/mL) for up to 24 hours resulted in a time-dependant decline in cell viability. This decrease was greatest in cells stored at 37°C (62% viability) followed by 24°C (70% viability) and 4°C (81% viability). Those cells stored at 4°C for 24 hours maintained their capacity for osteogenic differentiation and new bone formation upon implantation into rats (Muraki et al., 2006). This early study suggested that this could be a simple, cost-effective route for the short-term distribution of therapeutic cells, avoiding the requirement for cryo-shipping. Indeed, the ability to use a simple buffered solution for hypothermic storage, exploiting the low metabolic demand of cells in "suspended animation", could be a direct route towards a cGMP-compliant CTP.

Later studies demonstrated that BM-hMSCs produced under cGMP conditions stored at 20×10^6 cells/mL in a 5 mL syringe at 4°C for 18 hours maintained greater than 80% viability in normal saline, with no discernable difference between different storage media: maintenance medium (α-MEM with platelet lysate) or saline with human serum albumin (HSA). Approximately 55% of cells stored in saline subsequently attached to 3D scaffolds within 1 hour, which was not improved through the addition of HSA but was greater for those stored in maintenance medium. Although proliferation was greater for cells stored in maintenance medium or saline with HSA, their capacity for osteogenic differentiation and bone-forming capacity when implanted into mice were comparable (Veronesi et al., 2014). Conversely, BM-hMSCs suspended in saline (2×10^6 cells/mL) at 4°C exhibited a rapid decline in viable cell number over a 72-hour period, with attenuation of growth post-storage. Whilst this observation was abrogated through the addition of human serum albumin (HSA) from specific manufacturers, growth, albeit improved, was still less than 40% of that observed in freshly isolated cells (Lane et al., 2009), demonstrating the importance of examining cell function following storage. In support of this are more recent studies where BM-hMSCs suspended in saline (5×10^6 cells/mL) at either room temperature or 4°C, whilst demonstrating only a slight decrease in viability over a 6-hour period, formed considerably fewer colonies and their trilineage differentiation capacity was much diminished (Sohn et al., 2013) suggesting that, even though cells stored for a short period of time may appear viable, their functionality may be severely diminished. The results posed by Lane et al. (2009), that suggest the effects of HSA could be manufacturer independent, may go some way to explain the variable outcomes with regards to its supplementation in storage media.

Two recent reports, however, suggest that HSA, accompanied by glucose, may be important for the storage of human adipose-derived MSCs (hASCs) over periods of time up to two and a half days (Galvez et al., 2014, Galvez-Martin et al., 2014). From these studies, it does appear feasible to store suspended cells in simple buffer solutions, possibly requiring HSA or platelet lysate supplementation, for maintaining the viability of suspended cells over a short-term storage period. Sub-optimal performance of MSCs following release from storage, however, is highlighted, potentially being initiated relatively early.

Whilst some evidence does suggest that MSC requirements in terms of storage media are low, much effort has been spent in optimising media formulations for longer storage periods. The availability of commercial media formulations for the cold storage of cells has led to their implementation in MSC preservation. Subconfluent, adherent cells stored in HypoThermosol-FRS® at 4°C for 2 and 4 days have been demonstrated to yield 100 and 85% viable cell recovery respectively, when compared to cells kept at 37°C in culture. Recovered cells had normal proliferation, MSC marker expression and osteogenic potential (Ginis et al., 2012). This is in line with data presented on the manufacturer's website (BioLife Solutions, 2015), where the percentage recovery of viable cells after 5 days was over 80%, although this dropped to under 40% after 7 days, demonstrating limitations in terms of storage time. This limit is potentially prohibitory, as a preservation time of between 7 and 10 days may be required for a cellular therapy to allow time for quality, safety and efficacy testing before transportation and administration to the patient (Robinson et al., 2014), and even longer for (USP)-sterility testing (Naughton, 2002). The longest storage period demonstrated has been up to 3 weeks in a buffer and media solution containing 40 mM trehalose, a sugar more commonly exploited for its cryoprotective effects (Eroglu et al., 2000).

The authors demonstrated that human umbilical cord MSCs maintained at 4°C at a density of 2×10^6 cells/mL had a viability of over 90% after 7 days in the presence of 40 mM trehalose, whilst in its absence, viability dropped to under 30%. When the storage period was increased to 2 weeks, cell viability was maintained above 80% and, by 3 weeks, above 70%. It is worth mentioning, however, that the number of cells staining negative for propidium iodide (PI) and annexin V over the time course decreased, with less than 45% negative cells by 3 weeks suggesting the accumulation of apoptotic cells. Nevertheless, the viability of cells without trehalose declined rapidly, suggesting its usefulness in hypothermic preservation for relatively long periods of time (Di et al., 2012).

It is interesting that the studies discussed so far have concentrated on cold storage (e.g. 0–5°C), whilst recent evidence suggests that MSCs are susceptible to cold shock. A marked decrease in the viability of adherent MSC monolayers was observed by Corwin et al. (2011) following 24 to 48 hours at 4°C, even in commercially available Viaspan®, although this storage media did preserve viability to a degree. The reduction in viability was associated with increased apoptosis and necrosis with an increased expression of pro-apoptotic proteins, decreased expression of the pro-survival protein Bcl2, and increased endoplasmic reticulum stress and unfolded protein response (UPR) stress pathway activation. Cells stored for 48 hours did however demonstrate a return in the level of viability close to that of control following 48 hours in culture in the presence of the polyphenol resveratrol (Corwin et al., 2014). Although no differences between room temperature and 4°C storage were observed with BM-MSCs suspended in saline over a 6-hour period (Sohn et al., 2013), temperature was found to have a substantial effect on the preservation of MSCs cultured on fibrin microbeads.

Cells cultured on fibrin microbeads, sealed in cryovials containing medium, and stored at room temperature (~24°C) for 6 days resulted in almost 100% viable cell recovery, whilst almost 80% of cells died at 4°C. Those cells stored at room temperature maintained MSC marker expression, were able to proliferate, and expressed markers of osteo- and chondrogenic lineages upon induction (Gorodetsky et al., 2011). Indeed, later studies suggested that hMSCs encapsulated in a calcium alginate hydrogel maintained over 80% viability following 5 days ambient storage, a yield comparable with cryopreservation over the same period (Chen et al., 2013). Contrary to this, the short-term storage of hASCs over two and a half days in supplemented Lactate Ringer's solution suggested that 4°C storage was preferable for survival when compared to 8°C and 25°C, with 37°C being highly detrimental to cell viability (Galvez-Martin et al., 2014).

In terms of temperature, the detrimental effects at 37°C offer the only clue to what is optimal for storage. This is illustrated in Table 7.1, where there is a limited recovery of suspended viable cells at short- to mid-storage times when exposed to 37°C. This is unsurprising as, within a normothermic environment, cells will be actively metabolising and thus, will have a requirement for oxygen and nutrients, and will be sensing a sub-optimal environment for their survival. In terms of hypothermic temperatures however, this is slightly less clear as there has been success with the storage of MSCs at both refrigerated (4°C) and more ambient (8–25°C) temperatures over a range of time-points (Table 7.1). Of relevance are the longer time-points (>4 days) where 4°C storage of adherent cells in the leading hypothermic storage medium, Hypothermosol-FRS®, results in a good level of viability up to 4–5 days, but appears to drop sharply at 7 days. Similarly, cells seeded on fibrin beads exhibit a dramatic decrease in viability at 6 days when stored at 4°C, but this is not the case for those stored at ambient temperature. It is possible that this displays the limitations of refrigerated storage of adherent cells.

In terms of cells in suspension, storage in calcium alginate at ambient temperatures results in a good level of viability after 5 days, but this has not been investigated at longer time-points allowing no comparison with suspended cells stored at 4°C in buffer supplemented with trehalose, which demonstrated greater than 80% viability for up to 2 weeks

Table 7.1 Parameters and cell viability outcomes for the short-term storage of MSCs

					% viability					
Ref.	**Time (days)**	**Temp. (°C)**	**Storage media**	**Ad./Sp.**	**≤20**	**>20, ≤40**	**>40, ≤60**	**>60, ≤80**	**>80**	**Comments**
[1]	0.25	4 RT	Saline	Sp.					*	↓ CFU-f, ↓Tdiff.
[2]	0.5	37	Lactated Ringer's	Sp.				*		
[3]	0.75	4	Saline/α-MEM wt. PL	Sp.					*	↓ attachment and growth normal
[4]	1	4	PBS	Sp.					*	Odiff. normal
[5]	1	4	Saline + HSA	Sp.					#	↓ growth post-storage
[6]	1	4	Viaspan	**Ad.**				#		Metabolic return w/time
[4]	1	24	PBS	Sp.				*		
[4]	1	37	PBS	Sp.				*		
[2]	1	37	Lactated Ringer's	Sp.			*			

(Continued)

Table 7.1 (Continued)

Ref.	Time (days)	Temp. (°C)	Storage media	Ad./Sp.	% viability ≤20	>20, ≤40	>40, ≤60	>60, ≤80	>80	Comments
[2]	1.5	8, 25	Lactated Ringer's	Sp.					*	
[2]	2	4	Lactated Ringer's	Sp.					*	
[5]	2	4	Saline + HSA	Sp.					#	↓ growth post-storage
[6]	2	4	HBSS wt. resveratrol	**Ad.**			#			Metabolic return w/time
[7]	2.25	4	Lactated Ringer's	Sp.					*	
[7]	2.5	4	Lactated Ringer's	Sp.				*		
[2]	2.5	4	Lactated Ringer's	Sp.				*		Ktype, MME, Adiff. normal
[2]	2.5	8, 25	Lactated Ringer's	Sp.				*		
[2]	2.5	37	Lactated Ringer's	Sp.		*				
[8]	4	4	Hypothermosol-FRS®	**Ad.**					#	MME & Odiff. normal
[9]	5	4	Hypothermosol-FRS®	**Ad.**					#	
[10]	5	18-22	Alginate, LG-DMEM	Sp.					#	MME and growth normal

[11]	6	~24	Fibrin bead, α-MEM	**Ad.**			♯	MME, growth, Odiff. and Cdiff. normal
[11]	6	4	Fibrin bead, α-MEM	**Ad.**	♯			
[9]	7	4	Hypothermosol-FRS®	**Ad.**	♯			
[12]	7	4	Buffer wt. trehalose	Sp.			*	
[12]	14	4	Buffer wt. trehalose	Sp.			*	
[12]	21	4	Buffer wt. trehalose	Sp.		*		46% annexin-V & PI +ve MME & growth normal

[1]Sohn *et al.*, 2013, [2]Galvez-Martin *et al.*, 2014, [3]Veronesi *et al.*, 2014, [4]Muraki *et al.*, 2006, [5]Lane *et al.*, 2009, [6]Corwin *et al.*, 2014, [7]Galvez *et al.*, 2014, [8]Ginis *et al.*, 2012, [9]BioLife Solutions, [10]Chen *et al.*, 2013, [11]Gorodetsky *et al.*, 2011, [12]Di *et al.*, 2012. Entries are listed in order of storage time and storage temperatures: 4°C, 8–25°C, and 37°C. Asterisks (*) indicate where membrane exclusion has been relied on to calculate the % viability indicated; hash tags (♯) indicate where viable cell counts relative to initial viable count, or measures of metabolism have been used to calculate the % viability indicated. Where different storage media have been examined within a study, the best outcome has been selected for inclusion. Abbreviations: **Ad.**, adherent cells; **Sp.**, suspended cells; **CFU-f**, colony-forming unit fibroblast potential, **Tdiff.**, trilineage differentiation potential; **Odiff.**, osteogenic differentiation potential; **Adiff.**, adipogenic differentiation potential; **Cdiff.**, chondrogenic differentiation potential; **MME**, MSC Marker Expression; **LG-DMEM**, Low-Glucose Dulbecco's Modified Eagle's Medium; **PI**, Propidium Iodide.

(Table 7.1). It is important to note, however, that this study, along with others listed in the table, has relied on membrane exclusion assays to gauge the relative viability of cells (indicated with an asterisk) without relating back to the starting viable cell population or using other methods such as comparing levels of metabolism after a period of recovery.

As deviations in osmotic pressure during storage cause cell membrane rupture and cell lysis, membrane exclusion assays are not sufficient to account for these dead cells, potentially overestimating cell viability. In order to reach conclusions over the best conditions for the storage of MSC-based CTPs, methods that reflect true cell viability and function need to be implemented to achieve a reliable measure of total functional cell recovery. Regardless of this, studies examining cells, both in suspension and adherent to a surface, represent a strong promise that hypothermic storage can be applied for the storage and distribution of MSCs, at least in the short term. Clues as to the optimal temperature for hypothermic storage, however, may lie in those studies carried out in other cell types.

7.2.2 Optimal Temperature for Cell Storage

Exposing cells to hypothermic conditions as a means to maintain them in suspended animation for storage has been examined for a number of cell types, ranging from transformed cell lines to more sensitive primary cells and tissues. After a 1-week storage period of adherent retinal pigment epithelial cells in HEPES-buffered MEM at a range of temperatures between 4 and 37°C, Pasovic et al. (2013) found the optimal recovery of viable cells at 16°C (~50%). At temperatures either above or below the ranges of 12–20°C, there was a considerable drop in recovery, with 5°C yielding less than 5% recovery, with corresponding alterations in cell morphology and phenotype (Pasovic et al., 2013). Similarly, Chinese hamster ovary (CHO) cells stored in suspension for 4 days revealed 17°C as an optimal temperature for cell viability (>80%), with 4°C storage resulting in decrease in viability to 20–30%, also attenuating their growth following release from storage (Hunt et al., 2005).

The effect of temperature on CHO cell storage was later supported by Roobol et al. (2009). where CHO, P19 embryonic carcinoma, and NIH 3T3 fibroblast cell lines were stored at as adherent monolayers. Cells maintained at 27°C were morphologically indistinguishable from those maintained at 37°C for 6 hours; however, exposure of cells to 4°C resulted in rounding, microarchitectural disruption and severe attenuation of protein synthesis. Furthermore, organ-cultured corneal epithelium seeded onto amniotic membrane has demonstrated better preservation at 23–31°C after 1 week when compared to 5°C in Optisol-GS hypothermic storage medium (Raeder et al., 2007), although later studies revealed detachment of conjunctival epithelia from amniotic membrane started to occur after 4 days at 23°C (Eidet et al., 2012). Similar detachment from amniotic membrane, however, was observed with limbal epithelium stored at 4°C after 4 days (Liu et al., 2012), so this may not be a consequence of storage temperature. Ambient storage of human limbal epithelial cells encapsulated in calcium alginate discs have also proved to yield a greater viable recovery of cells compared to chilled storage over 7 days, as well as to form competent cultures post-release (Wright et al., 2012). In addition, the short-term storage of a number of recombinant cell lines at 23°C has proved suitable for "ready to use" cell-based assays (Wise et al., 2009). Collectively, less severe hypothermia may be more suitable for cell storage compared to cold storage, at least for certain cell types.

Conversely, haematopoietic progenitor cells obtained from cord blood, peripheral blood and bone marrow experienced greater colony forming unit (CFU) loss when stored for 72 hours at 20°C compared to 4°C (Fry et al., 2013). Indeed, the overnight storage of autologous stem cell apheresis at 2–6°C prior to processing has been routinely implemented in some hospitals, reducing the costs and complications involved with taking multiple apheresis samples whilst having no impact on viability or engraftment (Parkins et al., 2006). This is mirrored by the successful storage of various cell types at 4°C, including hepatocytes encapsulated in calcium alginate for 2 days (Mahler et al., 2003), corneal endothelial cells for 2 days (Corwin et al., 2011), hESC-derived cardiomyocyte organoid cultures for 3 days (Dai et al., 2007) and African green monkey kidney-derived BGM cells for 8 days (Moce-Llivina and Jofre, 2004), as well as human corneal limbal tissue for 8 days (Li et al., 2013).

Although certain studies may suggest refrigeration may be optimal for hypothermic cell storage, where the different temperatures have been directly compared, there are at least seven publications that reported that ambient (12–27°C) conditions were optimal across seven cell types including those for MSCs, transformed cell lines and corneal epithelial cells, one account where a temperature difference did not affect MSC viable recovery, and only three accounts for cold (0–5°C) conditions being optimal for MSC and haematopoietic progenitor storage. Therefore, cell preservation in milder (>12°C) hypothermic conditions may be a better choice for the transportation of therapeutic cells. This is reflected by the storage conditions employed by many approved CTPs, including ChondroCelect® (15–22°C), Epicel® (13–23°C) and Apligraf® (20–23°C), demonstrating the usefulness of non-chilled hypothermic conditions for the storage and distribution of cell-based therapies.

7.3 The Application of Hypothermic Storage in Cell-based Therapies

There are a number of benefits that hypothermic liquid-state storage could convey when applied to the cell therapy logistical supply chain, extending the time window between the clinic and sites of manufacturing. The studies discussed have represented cells stored in various storage media, from simple physiological saline solutions to more complex formulations such as Hypothermosol-FRS®, in both suspended and adherent states, and at a range of temperatures, many with favourable outcomes which are encouraging. One of the major limitations in the outcomes presented, at least in the case of MSCs, are the storage periods that have been investigated, with only a single study (Di et al., 2012) demonstrating successful storage past 1 week, and the leading hypothermic preservation media Hypothermosol-FRS® demonstrating limitations at this 7-day time-point (BioLife Solutions, 2015). In order to realise the full potential benefit of hypothermic storage in clinical applications, longer periods should be investigated with comprehensive characterisation of cell viability, genetic stability and function. Longer periods will allow for full quality, safety, efficacy and sterility testing of CTPs prior to administration to patients. In addition, when considering the appropriate methods for the storage and distribution of living biologics, there are a number of parameters that require consideration. These include packaging, suitable for maintaining appropriate conditions for the CTP and for cell delivery, the effect of temperature transitions, and the effect of physical stresses on cells incurred as a result of transportation. Importantly, as well as cells being viable, any cell therapy product must be assessed for physiochemical characteristics, biological function, potency, biocompatibility, stability, consistency and traceability (Bravery et al., 2013; Trainor et al., 2014). Thus, a standard protocol for cell storage and distribution that takes into account all of these parameters is of utmost importance for the widespread implementation of hypothermic preservation.

It is also important to consider how hypothermic storage could be integrated with cryogenic storage for application in allogeneic therapies where a short time window exists following cell thawing. It is encouraging that certain approaches demonstrated to promote a suitable outcome for hypothermic preservation, have also been reported to show benefits in cryogenic preservation, including encapsulation of cells in alginate (Mahler et al., 2003; Malpique et al., 2010, Serra et al., 2011). Furthermore, alginate has also been demonstrated as useful for the scalable expansion of embryonic stem cells (Serra et al., 2011). From a bioprocessing point of view, a system that is able to integrate upstream expansion, as well as downstream cryopreservation and/or hypothermic storage could be key for the production of a scalable, cGMP-compliant CTP.

7.4 Concluding Remarks

Acceleration in the application of cell-based therapies, both in late phases of clinical trial and in the approval of CTPs, accompanied by the complex logistical platform involved with their manufacture, quality assurance and clinical delivery, warrants considerable attention into methods to preserve cells during storage and distribution. This chapter has outlined hypothermic preservation as a potential method to achieve this, and puts forwards a number of scenarios where its implementation could be of huge benefit within the cell therapy industry. There have been considerable advances in the research field as to the best approaches for hypothermic cell storage, and it is anticipated that this will develop further in order to facilitate a robust, scalable approach for the extended preservation of cells suitable for adoption within cell therapy applications.

References

Al-Fageeh, M.B. and Smales, C.M. (2006) Control and regulation of the cellular responses to cold shock: The responses in yeast and mammalian systems. *Biochem. J.*, **397**, 247–59.

Apligraf: APG Factsheet, last updated June 2011 and Prescribing Information, last updated December 2010. Available at: http://www.apligraf.com

Baust, J.G. (2006) Concepts in biopreservation, in *Advances in Biopreservation* (eds J.G. Baust and J.M. Baust), CRC Press, Boca Raton, FL, pp. 1–14.

Biolife Solutions 2015. *HypoThermosol Performance Evidence* [Online]. Available at: http://biolifesolutions.com/evidence. Accessed 14 September 2016.

Bissoyi, A., Nayak, B., Pramanik, K. and Sarangi, S.K. (2014) Targeting cryopreservation-induced cell death: A review. *Biopreservation and Biobanking*, **12**, 23–34.

Bravery, C.A., Carmen, J., Fong, T. *et al.* (2013) Potency assay development for cellular therapy products: an ISCT review of the requirements and experiences in the industry. *Cytotherapy*, **15**, 9–19.

Campbell, L.H., Taylor, M.J. and Brockbank, K.G.M. (2013) Development of pancreas storage solutions: Initial screening of cytoprotective supplements for beta-cell survival and metabolic status after hypothermic storage. *Biopreservation and Biobanking*, **11**, 12–18.

Carticel: Prescribing Information, Last updated June 2010. Available at: http://www.carticel.com

Chen, B., Wright, B., Sahoo, R. and Connon, C.J. (2013) A novel alternative to cryopreservation for the short-term storage of stem cells for use in cell therapy using alginate encapsulation. *Tissue Engineering Part C Methods*, **19**, 568–76.

Chondrocelect: EPAR-product information, last updated 20 November 2013. Available at: http://www.ema.europa.eu

Corwin, W.L., Baust, J.M., Baust, J.G. and Van Buskirk, R.G. (2011) The unfolded protein response in human corneal endothelial cells following hypothermic storage: Implications of a novel stress pathway. *Cryobiology*, **63**, 46–55.

Corwin, W.L., Baust, J.M., Baust, J.G. and Van Buskirk, R.G. (2014) Characterization and modulation of human mesenchymal stem cell stress pathway response following hypothermic storage. *Cryobiology*, **68**, 215–26.

Dai, W., Field, L.J., Rubart, M. *et al.* (2007) Survival and maturation of human embryonic stem cell-derived cardiomyocytes in rat hearts. *J. Mol. Cell Cardiol.*, **43**, 504–16.

Deb, K. (2009) A sweet potion to put embryonic stem cells to sleep. *Scientific World Journal*, **9**, 236–49.

Dermagraft: Directions for use, 2013. Available at: http://www.dermagraft.com

Di, G.H., Wang, J.X., Liu, M.X. *et al.* (2012) Development and evaluation of a trehalose-contained solution formula to preserve hUC-MSCs at 4 degrees C. *Journal of Cellular Physiology*, **227**, 879–84.

Eidet, J.R., Utheim, O.A., Raeder, S. *et al.* (2012) Effects of serum-free storage on morphology, phenotype, and viability of *ex vivo* cultured human conjunctival epithelium. *Exp. Eye Res.*, **94**, 109–16.

Epicel: Directions for use. Last updated 2014. Available at: http://www.epicel.com

Eroglu, A., Russo, M.J., Bieganski, R. *et al.* (2000) Intracellular trehalose improves the survival of cryopreserved mammalian cells. *Nature Biotechnology*, **18**, 163–7.

Foley, L. and Whitaker, M. (2012) Concise review: Cell therapies – The route to widespread adoption. *Stem Cells Transl. Med.*, **1**, 438–47.

Fry, L.J., Giner, S.Q., Gomez, S.G. *et al.* (2013) Avoiding room temperature storage and delayed cryopreservation provide better post-thaw potency in hematopoietic progenitor cell grafts. *Transfusion*, **53**, 1834–42.

Galvez, P., Martin, M.J., Calpena, A.C. *et al.* (2014) Enhancing effect of glucose microspheres in the viability of human mesenchymal stem cell suspensions for clinical administration. *Pharmaceutical Research*, **31**, 3515–28.

Galvez-Martin, P., Hmadcha, A., Soria, B. *et al.* (2014) Study of the stability of packaging and storage conditions of human mesenchymal stem cell for intra-arterial clinical application in patient with critical limb ischemia. *European Journal of Pharmaceutics and Biopharmaceutics*, **86**, 459–68.

Ginis, I., Grinblat, B. and Shirvan, M.H. (2012) Evaluation of bone marrow-derived mesenchymal stem cells after cryopreservation and hypothermic storage in clinically safe medium. *Tissue Engineering Part C Methods*, **18**, 453–63.

Gorodetsky, R., Levdansky, L., Gaberman, E. *et al.* (2011) Fibrin microbeads loaded with mesenchymal cells support their long-term survival while sealed at room temperature. *Tissue Engineering Part C Methods*, **17**, 745–55.

Hernon, C.A., Dawson, R.A., Freedlander, E. *et al.* (2006) Clinical experience using cultured epithelial autografts leads to an alternative methodology for transferring skin cells from the laboratory to the patient. *Regenerative Medicine*, **1**, 809–21.

Hourd, P., Ginty, P., Chandra, A. and Williams, D.J. (2014) Manufacturing models permitting roll out/scale out of clinically led autologous cell therapies: Regulatory and scientific challenges for comparability. *Cytotherapy*, **16**, 1033–47.

Hunt, L., Hacker, D.L., Grosjean, F. *et al.* (2005) Low-temperature pausing of cultivated mammalian cells. *Biotechnol. Bioeng.*, **89**, 157–63.

Kemp, P. (2006) History of regenerative medicine: Looking backwards to move forwards. *Regenerative Medicine*, **1**, 653–69.

Lane, T.A., Garls, D., Mackintosh, E. *et al.* (2009) Liquid storage of marrow stromal cells. *Transfusion*, **49**, 1471–81.

Li, C., Dong, N., Wu, H. *et al.* (2013) A novel method for preservation of human corneal limbal tissue. *Invest. Ophthalmol. Vis. Sci.*, **54**, 4041–7.

Li, M.D., Atkins, H. and Bubela, T. (2014) The global landscape of stem cell clinical trials. *Regenerative Medicine*, **9**, 27–39.

Liu, T., Wang, Y., Duan, H.Y. *et al.* (2012) Effects of preservation time on proliferative potential of human limbal stem/progenitor cells. *International Journal of Ophthalmology*, **5**, 549–54.

Loi, P., Iuso, D., Czernik, M. *et al.* (2013) Towards storage of cells and gametes in dry form. *Trends Biotechnol.*, **31**, 688–95.

Mahler, S., Desille, M., Fremond, B. *et al.* (2003) Hypothermic storage and cryopreservation of hepatocytes: The protective effect of alginate gel against cell damages. *Cell Transplantation*, **12**, 579–92.

Malpique, R., Osorio, L.M., Ferreira, D.S. *et al.* (2010) Alginate encapsulation as a novel strategy for the cryopreservation of neurospheres. *Tissue Eng. Part C Methods*, **16**, 965–77.

Mannello, F. and Tonti, G.A. (2007) Concise review: No breakthroughs for human mesenchymal and embryonic stem cell culture: Conditioned medium, feeder layer, or feeder-free; medium with fetal calf serum, human serum, or enriched plasma; serum-free, serum replacement nonconditioned medium, or *ad hoc* formula? All that glitters is not gold!. *Stem Cells*, **25**, 1603–9.

Moce-Llivina, L. and Jofre, J. (2004) A method to maintain mammalian cells for days alive at 4 degrees C. *Cytotechnology*, **46**, 57–61.

Muraki, K., Hirose, M., Kotobuki, N. *et al.* (2006) Assessment of viability and osteogenic ability of human mesenchymal stem cells after being stored in suspension for clinical transplantation. *Tissue Engineering*, **12**, 1711–19.

Murphy, M.B., Moncivais, K. and Caplan, A.I. (2013) Mesenchymal stem cells: environmentally responsive therapeutics for regenerative medicine. *Experimental and Molecular Medicine*, **45**, e54.

Naughton, G.K. (2002) From lab bench to market: Critical issues in tissue engineering. *Ann. N.Y. Acad. Sci.*, **961**, 372–85.

Olinga, P., Merema, M., Slooff, M.J.H. *et al.* (1997) Influence of 48 hours of cold storage in University of Wisconsin organ preservation solution on metabolic capacity of rat hepatocytes. *Journal of Hepatology*, **27**, 738–43.

Pal, R., Hanwate, M. and Totey, S.M. (2008) Effect of holding time, temperature and different parenteral solutions on viability and functionality of adult bone marrow-derived mesenchymal stem cells before transplantation. *Journal of Tissue Engineering and Regenerative Medicine*, **2**, 436–44.

Pamphilon, D.H., Selogie, E. and Szczepiorkowski, Z.M. (2010) Transportation of cellular therapy products: report of a survey by the cellular therapies team of the Biomedical Excellence for Safer Transfusion (BEST) collaborative. *Vox Sang*, **99**, 168–73.

Parkins, M.D., Bahlis, N., Brown, C. *et al.* (2006) Overnight storage of autologous stem cell apheresis products before cryopreservation does not adversely impact early or long-term engraftment following transplantation. *Bone Marrow Transplantation*, **38**, 609–14.

Pasovic, L., Utheim, T.P., Maria, R. *et al.* (2013) Optimization of storage temperature for cultured ARPE-19 Cells. *Journal of Ophthalmology*, **2013**, Article ID 216359, 11 pages, doi: 10.1155/2013/216359

Pikula, M., Marek-Trzonkowska, N., Wardowska, A. *et al.* (2013) Adipose tissue-derived stem cells in clinical applications. *Expert Opin. Biol. Ther.*, **13**, 1357–70.

Provenge: Prescribing Information. Last updated 2011. Available from: http://www.provenge.com

Raeder, S., Utheim, T.P., Utheim, O.A. *et al.* (2007) Effects of organ culture and Optisol-GS storage on structural integrity, phenotypes, and apoptosis in cultured corneal epithelium. *Investigative Ophthalmology and Visual Science*, **48**, 5484–93.

Rhoads, L.S. (1994) Use of hypothermically stored keratinocyte cultures for *in-vitro* toxicology. *Journal of Toxicology-Cutaneous and Ocular Toxicology*, **13**, 231–48.

Robinson, N.J., Picken, and Coopman, K. 2014. Low temperature cell pausing: An alternative short-term preservation method for use in cell therapies including stem cell applications. *Biotechnology Letters*, **36**, 201–9.

Roobol, A., Carden, M.J., Newsam, R.J. and Smales, C.M. (2009) Biochemical insights into the mechanisms central to the response of mammalian cells to cold stress and subsequent rewarming. *Febs Journal*, **276**, 286–302.

Schulz, J.C., Germann, A., Kemp-Kamke, B. *et al.* (2012) Towards a xeno-free and fully chemically defined cryopreservation medium for maintaining viability, recovery, and antigen-specific functionality of PBMC during long-term storage. *Journal of Immunological Methods*, **382**, 24–31.

Seo, J.M., Sohn, M.Y., Suh, J.S. *et al.* (2011) Cryopreservation of amniotic fluid-derived stem cells using natural cryoprotectants and low concentrations of dimethylsulfoxide. *Cryobiology*, **62**, 167–173.

Serra, M., Correia, C., Malpique, R., Brito, C., Jensen, J., Bjorquist, P., *et al.* (2011) Microencapsulation technology: A powerful tool for integrating expansion and cryopreservation of human embryonic stem cells. *PLoS One*, **6**(8), e23212. doi: 10.1371/journal.pone.0023212

Sohn, H.S., Heo, J.S., Kim, H.S. *et al.* (2013) Duration of *in vitro* storage affects the key stem cell features of human bone marrow-derived mesenchymal stromal cells for clinical transplantation. *Cytotherapy*, **15**, 460–6.

Southard, J.H. and Belzer, F.O. (1995) Organ preservation. *Ann. Rev. Med.*, **46**, 235–47.

Swan, H., Virtue, R.W., Blount, S.G. Jr. and Kircher, L.T. Jr. (1955) Hypothermia in surgery; analysis of 100 clinical cases. *Ann. Surg.*, **142**, 382–400.

Syed, B.A. and Evans, J.B. (2013) Stem cell therapy market. *Nat. Rev. Drug Discov.*, **12**, 185–6.

Thirumala, S., Goebel, W.S. and Woods, E.J. (2013) Manufacturing and banking of mesenchymal stem cells. *Expert Opin. Biol. Ther.*, **13**, 673–91.

Trainor, N., Pietak, A. and Smith, T. (2014) Rethinking clinical delivery of adult stem cell therapies. *Nat. Biotechnol.*, **32**, 729–35.

Veronesi, E., Murgia, A., Caselli, A. *et al.* (2014) Transportation conditions for prompt use of *ex vivo* expanded and freshly harvested clinical-grade bone marrow mesenchymal stromal/stem cells for bone regeneration. *Tissue Engineering Part C Methods*, **20**, 239–51.

Wise, H., Abel, P.W. and Cawkill, D. (2009) Use of reduced temperature cell pausing to enhance flexibility of cell-based assays. *Journal of Biomolecular Screening*, **14**, 716–22.

Wright, B., Cave, R.A., Cook, J.P. *et al.* (2012) Enhanced viability of corneal epithelial cells for efficient transport/storage using a structurally modified calcium alginate hydrogel. *Regenerative Medicine*, **7**, 295–307.

Zeisberger, S.M., Schulz, J.C., Mairhofer, M. *et al.* (2011) Biological and physicochemical characterization of a serum- and xeno-free chemically defined cryopreservation procedure for adult human progenitor cells. *Cell Transplantation*, **20**, 1241–57.

8

Cell Therapy in Practice

Gustavo S. Figueiredo[1,3]*, Julian R. De Havilland*[2]*, Majlinda Lako*[3] *and Francisco C. Figueiredo*[1,3]

[1] *Department of Ophthalmology, Royal Victoria Infirmary, Newcastle upon Tyne, UK*
[2] *Cellular Therapies Facility, Newcastle University, Newcastle upon Tyne, UK*
[3] *Institute of Genetic Medicine, Newcastle University, Newcastle upon Tyne, UK*

8.1 Introduction

Advanced Therapy Medicinal Products (ATMPs) have been brought about by progress in scientific developments across a range of academic disciplines. As new experimental products developed based on genes (gene therapy), cells (cell therapy) and tissues (tissue engineering), challenges as to how these new products could be controlled and regulated to ensure public safety and efficacy, emerged. To cope with these developments within the European Union, Regulation (EC) No. 1394/2007 of the European Parliament and of the Council, on Advanced Therapy Medicinal Products was adopted in 2007 (Regulation (EC), 2007).

These regulations were introduced for two purposes; firstly to protect human health and secondly to enable the free movement of ATMPs within the European Union. There was recognition within the new regulations that ATMPs are different in their routes to development; more likely to be stemming from academic institutions, spin-outs and small- or medium-sized enterprises, than from major pharmaceutical companies – at least in the early stages of new ATMP development. Such organisations developing very specific gene, cell or tissue based therapies and focusing on single indications were seen as originators of these innovations and without specific provisions how these therapies could be developed into marketed medicines given the usual time and

Bioprocessing for Cell Based Therapies, First Edition. Edited by Che J. Connon.

costs of such ventures needed to be addressed if there was to be innovation in medical interventions for a wide range of indications.

The Regulations imposed strict duties on the originators of ATMPs to manufacture such therapies in accordance with Good Manufacturing Practice (GMP). These are a set of practice guidelines for the manufacturing, testing and quality assurance of a drug, in order to ensure that it is safe for human consumption. The World Health Organization (WHO) version of GMP has been adopted by regulators, and the pharmaceutical industry worldwide – particularly so in developing countries. EU-GMP has developed very similar requirements to WHO GMP, as have the FDA in the US. In the UK, the Medicines Act (1968), subsequently revised by The Human Medicines Regulations 2012 [SI 2012/1916], covered most aspects of GMP in what is officially known as Rules and Guidance for Pharmaceutical Manufacturers and Distributors. The Pharmaceutical Press (2015), commonly referred to as "The Orange Guide", named because of the colour of its cover.

What has been absolutely key to how gene, cell and tissue therapies are regulated across all of the stakeholder bodies is the classification of each individual new therapy. This is a very critical consideration and an understanding of that classification very early in the development of any of these new treatment modalities clearly has the potential to support efficient routes to market of what are very complex new medicines.

8.2 The Classification of ATMPs

Precise legal definitions for ATMPs are provided by the ATMP Regulations and EU Directive 2001/83/EC: Annex I Part IV (Directive 2001/83/EC (2001). This EU Directive provides the basic framework for the regulation of the manufacture, marketing and distribution of medicines for human use. It was incorporated into UK law by the Human Medicines Regulations (2012). There is therefore an EU-wide Community Code designed to bring harmonisation across the EU to the provision of medicines – including ATMPs.

According to Article 2(1)(a) of the ATMP Regulations, an *Advanced Therapy Medicinal Product* refers to:

1) a gene therapy medicinal product, as defined in Part IV of Annex I to Directive 2001/83/EC;

2) a somatic cell therapy medicinal product, as defined in Part IV of Annex I to Directive 2001/83/EC; and
3) a tissue engineered product, as defined in Article 2(1)(b) of Regulation (EC) No. 1394/2007.

In addition, Article (2)(1)(d) of the ATMP Regulations also gave a definition of *Combined ATMP*. These products cross the boundaries of definitions where the ATMP element is administered as part of a medical device.

Classifications of gene therapies are beyond the scope of this chapter. However, a somatic cell therapy product as defined by the ATMP Regulations refers to a biological medicinal product which fulfils the following two characteristics:

1) contains or consists of cells or tissues that have been subject to substantial manipulation so that biological characteristics, physiological functions or structural properties relevant for the intended clinical use have been altered, or of cells or tissues that are not intended to be used for the same essential function(s) in the recipient and the donor; and
2) is presented as having properties for, or is used in or administered to human beings with a view to treating, preventing or diagnosing a disease through the pharmacological, immunological or metabolic action of its cells or tissues.

For a cell therapy to be classified as an ATMP, a lot will depend on whether the cells are intended for homologous use, a definition of which can be found in the US FDA Code of Federal Regulations Title 21.3(c) (2014):

> Homologous use means the repair, reconstruction, replacement or supplementation of a recipient's cells or tissues with an HCT/P that performs the same basic function or functions in the recipient as in the donor.

And in Part (d) these regulations define HCT/P as:

> Human cells, tissues, or cellular or tissue-based products (HCT/Ps) means articles containing or consisting of human cells or tissues that are intended for implantation, transplantation, infusion, or transfer into a human recipient.

In addition, the interpretation of what amounts to "substantial manipulation" has significant impact on the classification of cell therapies as ATMPs. There is no definition of what amounts to *substantial* but to assist interpretation, Annex I to Regulation (EC) No. 1394/2007, specifies that: "cutting, grinding, shaping, centrifugation, soaking in antibiotic or antimicrobial solutions, sterilization, irradiation, cell separation, concentration or purification, filtering, lyophilisation, freezing, cryopreservation, and vitrification", shall not be considered as substantial manipulations. There appears to be little or no guidance as to how combinations of these actions could affect ATMP classification, but it is clear that decisions by regulators will be made on a case-by-case basis, assessing the totality of manipulations that are applied, together with the intended use of the cell based product.

The list provided in the ATMP Regulations though is non-exhaustive and the threshold as to whether a cell therapy product is the result of substantial manipulation may sometimes be subjective. It is therefore accepted that the complex nature of cell therapies may result in some forms of therapeutic product receiving conflicting opinions as to classification – or even closely related processes being alternatively classed as a result of very subtle differences in process protocols.

It should also be remembered that definitions and regulations may change in the course of new drug development, which itself will take years. Changes to regulations could impact on the classification of a cell therapy product partway in the course of its route to market. Indeed, in this regard the European Medicines Agency is actively consulting on the definition of "substantial manipulation" (Qi et al., (2009), and where the axe eventually falls clearly has an impact on cell therapies in current pipelines. Similarly, there may be an impact on current cell products that, until now, have fallen outside of ATMP classification, but which are clearly exposed to complex protocols in their preparation, for example, the isolation of autologous and allogeneic pancreatic Islets of Langerhans (Figure 8.1).

The processing of pancreatic islets does provide a good example when considering the classification of a cell therapy product. The protocol for islet isolation is complex, involving the enzymatic dissociation of the tissue, often coupled to purification of islets over a density gradient, and then a period of storage in culture medium pending final product packaging, product release and eventual treatment of a named patient (Qi et al., 2009). The EU legislation applies to all elements of the production process and yet the preparation of islets is not classed

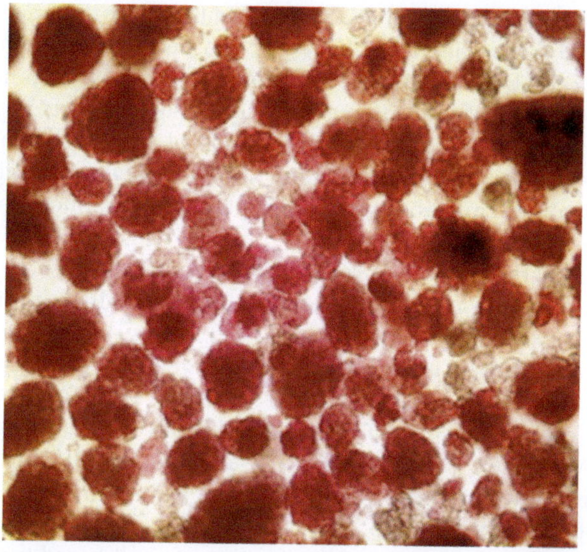

Figure 8.1 Purified human Islets of Langerhans as a cell therapy product.

as an ATMP. The cells are deemed for homologous use and despite the complexity of the protocol, are viewed as not substantially manipulated (EMA/CAT, 2010). It is believed there is no change in the biological characteristics of the islets and so they are not classed as an ATMP.

The advantage to manufacturing in not classifying the product as an ATMP, but is that the product then falls under the EU Tissues and Cells Directive (2004). This EU Directive was carried into UK law by the Human Tissue (Quality and Safety for Human application) Regulations (2007), and it incorporated the EU Parent Directive (2004), together with two further EU Technical Directives (1965, 2004) that encompassed the setting of standards of quality and safety for the donation, procurement, testing, processing, preservation, storage and distribution of human tissues and cells. Manufacturing products under this heading comes under the auspices of the Human Tissue Authority in the UK. Created by the Human Tissue Act (2004), its objectives are to:

> ...be the regulating authority for matters relating to activities such as anatomical and post-mortem examinations, transplantations and the storage of human material for education, training and research.

The HTA also acts as the UK competent authority under the EU Tissue and Cells Directives. As such, the inspectorate arm of this branch of the Department of Health is as rigorous, and some would say as powerful, as the Medicines and Healthcare products Regulatory Agency (MHRA). Though the obligations of the HTA appear less demanding, processing of cells under an HTA licence is still highly regulated and permits their use as transplantation products, for example, bone marrow products or Islets of Langerhans.

Ultimately though, the purpose of ATMP regulations is to harmonise the EU-wide route to market of commercially exploited cell therapies as medicines by providing a framework for establishing safety and efficacy. The significant advantage of ATMP status is that the therapy can then be, subject to approval, offered into clinical trials. If successful, such a therapy could be given a market authorisation (MA) to be sold across the EU. Without ATMP status, commercialisation of a cell-based product as a medicine for EU-wide distribution is not possible.

It is clear therefore that the classification of a cell-based therapy is a very significant element in decision-making as to the routes taken for development of each potential new cellular therapeutic (Figure 8.2). The flow chart in the figure details the principal questions to consider when seeking to classify a cell therapy product as an ATMP or as something that does not qualify to be classed as an ATMP.

8.3 European Regulations

The ATMP regulations are technically an amendment to EU Directive 2001/83/EC. This EU Directive was adopted to provide EU-wide consistency for the governance of medicinal products for human use, with the duty to safeguard public health at its core, but with the removal of trade barriers and national inconsistencies in regulations as a secondary objective. At the heart of the Directive is the mandatory requirement for what is known as a "Manufacturing Authorisation". This defines the terms under which a medicinal product is authorised in the EU. The MA as issued by a relevant authority provides permission to market a medicinal product, and includes a technical dossier with data submitted by the applicant in accordance with Articles 8(3) to 11 of Directive 2001/83/EC and Annex I, Articles 6(2) and 31(2) of Regulation (EC) No 726/2004, or Article 7 of Regulation (EC) No 1394/2007.

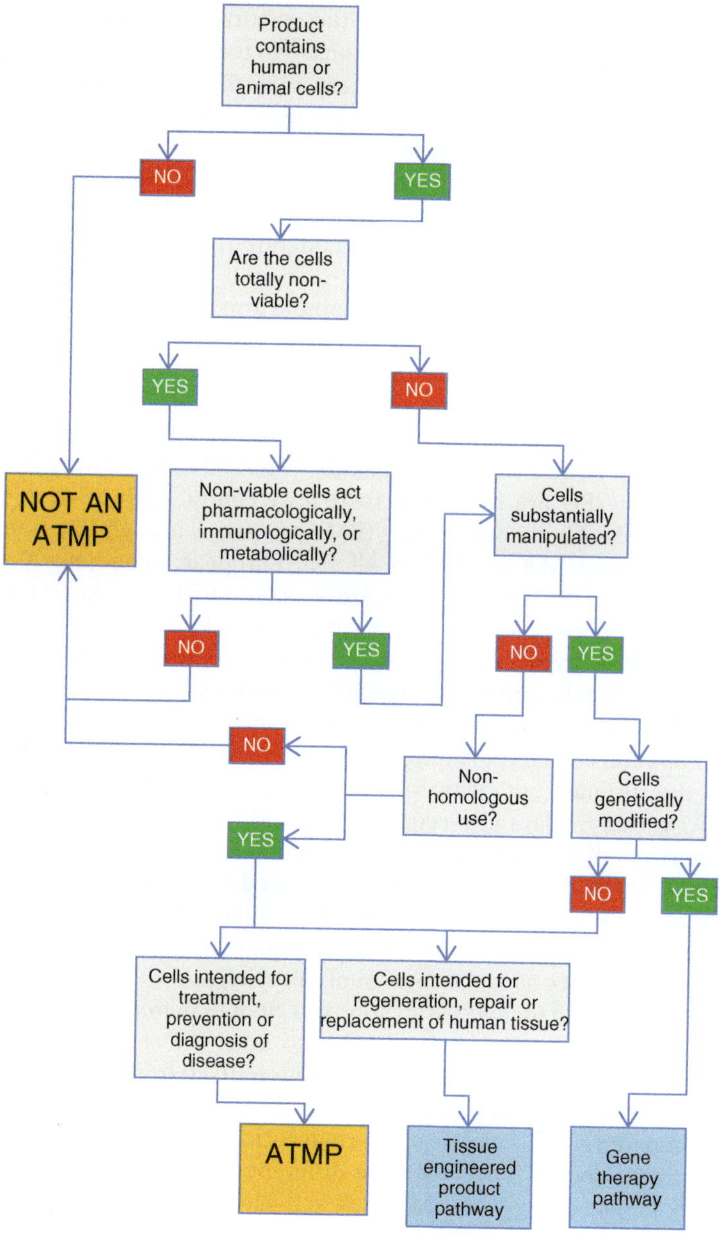

Figure 8.2 Decision tree for the classification of cell therapy medicinal products.

The origins of the MA stem from the reaction to the Thalidomide tragedy in the 1960s, when babies exposed to the drug *in-utero* developed deformities. Dated the 26 January 1965, Commission Directive 65/66/EEC/1 (1965) was the first European pharmaceutical directive aimed at setting standards for the approval of medicines. Now there is a body of regulation collectively known as "EudraLex" and consisting of 10 volumes governing medicinal products for human and veterinary use in the European Union (http://ec.europa.eu/health/documents/eudralex/index_en.htm).

The MA is in effect a licence to manufacture and commercialise a medicinal product for human use. To obtain an MA for the manufacture of licensed medicinal products, an application must be made to a competent authority. There are variations in application process dependent on circumstance, but in light of the specialist nature of ATMPs, by Article 3(1) of Regulation 726/2004, all ATMPs marketed in the EU are mandated to require central authorisation by the European Medicines Agency (EMA) (2004). Through this route, the centralised MA is valid in all EU, European Economic Area, and European Free Trade Association states that include Iceland, Liechtenstein and Norway. Upon submission of a valid application, the evaluation by the Committee for Medicinal Products for Human Use (CHMP) takes up to 210 days before they issue a scientific opinion as to whether the medicinal product may be authorised or not. This opinion is then transmitted to the European Commission, which has the ultimate authority for granting a marketing authorisation (MA) within 67 days of receipt of the CHMP opinion. This route to authorisation must also follow the Common Technical Document (CTD). The CTD is an internationally agreed format for the preparation of applications to be submitted to regulatory authorities in the three ICH regions of Europe, USA and Japan. Though the CTD provides no guidance as to the content of a dossier, nor which studies and data are required for successful approval., it does define the appropriate format for data in an application to meet the standards of the International Conference on Harmonisation (ICH) framework agreed in 2000, with agreed revisions ever since.

Compared to the classical pharmaceutical route to market, the MA requirements for ATMPs are different. Whilst seeking to safeguard public safety, the EMA has actively sought to support the innovators of this new class of medicine. The main aspects of ATMP regulation remain the centralised MA. Within that are specific elements designed to facilitate ATMP development. These are the Hospital Exemption ruling and provision for the manufacture of ATMPs under a compassionate use or "Specials" procedure in the UK, the establishment of the Committee for Advanced Therapies (CAT), and financial incentives for small- and medium-sized enterprises (SMEs) to reduce the costs of navigating the licensing process.

8.3.1 Hospital Exemption (HE) and "Specials" Manufacturing

Manufacture under Hospital Exemption rules or the Specials procedures in the UK offers pathways to manufacture ATMPs that bypass the centralised marketing authorisation procedure. The HE provision at Article 3(7) of Directive 2001/83/EC was included in the EU regulations in recognition of the small scale and developmental nature of cell-related activities within hospitals. The provision at Article 5(1) of Directive 2001/83/EC allow that:

> ...a Member State may, in accordance with legislation in force and to fulfil special needs, exclude from the provisions of this Directive medicinal products supplied in response to a bona fide unsolicited order, formulated in accordance with the specifications of an authorised healthcare professional and for use by an individual patient under his direct personal responsibility.

HE applies to ATMPs which are prepared on a non-routine basis, according to specific quality standards, and used within the same member state, and in a hospital in accordance with a medical prescription for an individual patient. This is then similar to the *Magisterial Formula* for medicinal products prepared to a specific prescription. Specials manufacture can be within the UK, or a product imported to the UK, and is subject to a "special needs" test. This is interpreted to mean there is no other pharmaceutically equivalent licensed medicinal product. Whereas a product manufactured under HE rules must be administered in a hospital, a Specials product has no defined location for treatment.

Manufacture under the Hospital Exemption provisions (or Specials in the UK), still needs a manufacturing licence from an EU Member State's competent authority. The process is still subject to licence and inspection, if not the full ethical review process of a clinical trial, but HE and Specials potentially allow the use of ATMPs in very rare indications that might otherwise never be subjected to efficacious ATMP-type therapies. In seeking to permit HE and Specials, regulators will consider whether the manufacturing process can be considered "non-routine", a potentially subjective parameter against which to seek a definition. Early consultation and engagement with competent authorities is as critical in classifying the manufacturing provisions as may apply, as it is in classifying the potential product.

To date, only a few member states have taken a proactive role and published guidance for the Hospital Exemption and Specials. These include the UK (MHRA) and the Netherlands (Dutch Health Care Inspectorate). However, neither the Hospital Exemption nor the Specials pathways are routes for taking ATMPs to the point of retail distribution without an MA. EU Competent Authorities permitting HE and Specials must still ensure adequate standards of pharmacovigilance and traceability. In addition, specific quality standards for ATMPs under the Hospital Exemption and Specials are equivalent to those for which the EMA's centralised procedure would be mandatory. These include, but are not limited to, manufacturing to standards of GMP, though a Qualified Person is not mandatory. It must also be remembered that neither Hospital Exemption nor Specials are intended for use in clinical trials. Treatments of patients under these exemptions are to be considered as case studies. As such, data accrued from such case studies does not replace the need for properly conducted clinical trials should application for a centralised MA be considered.

8.3.2 Orphan Medicinal Product Designation

In December 1999, the European Parliament adopted Regulation (EC) No. 141/2000 (The Orphan Regulation, 1999). Orphan designation can be assigned to medicines under development for rare diseases, and is granted by the European Commission, subsequent to an initial opinion being made by the EMA's Committee for Orphan Medicinal Products.

Specific criteria must apply for the EMA to consider. These are:

- that it must be intended for the treatment, prevention or diagnosis of a disease that is life-threatening or chronically debilitating;
- that it exhibits a prevalence in the EU of not more than 5 in 10,000, or it must be unlikely that the medicine could generate sufficient returns to justify the investment required for its development;

- no satisfactory method of diagnosis, prevention or treatment of the condition is already authorised, or, if such a method does already exists, the medicine must be of significant benefit to those affected by the condition.

The designation confers incentives to trial sponsors to develop medicines that would otherwise fail a test of pure economic viability. These incentives include discounts on fees, and protocol assistance. Protocol Assistance is a form of scientific advice for designated orphan medicines that helps sponsors in addressing the EMA's requirements for a successful MA application. The Committee for Medicinal Products would consider such an MA application for Human Use. Orphan status for a licensed ATMP would attract a 10-year market exclusivity within the EU, but this can be extended to 12 years if the sponsor included clinical results from a paediatric investigation plan at the time of MA application.

8.3.3 Committee for Advanced Therapies

The Committee for Advanced Therapies (CAT) was established by the ATMP Regulations (Regulation (EC) No. 1394/2007). It is comprised of multidisciplinary scientific experts representing all EU member states, as well as patient and medical associations. The CAT has the remit to prepare a draft opinion on each ATMP application submitted to the European Medicines Agency, before the Committee for Medicinal Products for Human Use of the EMA adopts a final opinion on the granting of a marketing authorisation. The CAT also provides scientific advice. The scientific advice procedure provides a mechanism by which early opinion on ATMP classification, as well as on the required and proposed preclinical and clinical development programmes, can be assessed in order to support the MA application process. Critical of course to successful MA application for an ATMP is the demonstration of controlled, safe and consistent manufacturing, so-called Good Manufacturing Practice.

8.3.4 Good Manufacturing Practice (GMP)

Good Manufacturing Practice is defined, as:

> ...a system for ensuring that products are consistently produced and controlled according to quality standards. It is designed to

minimise the risks involved in any pharmaceutical production that cannot be eliminated through testing the final product.

The legal basis for EU GMP guidelines is art. 3, sub-section 2 of the European directive 2003/94/EC, in conjunction with art. 47, sub-section 2 of the European directive 2001/83/EC. The EU GMP guidelines are regularly updated to reflect scientific developments and progress. GMP covers all aspects of production from the starting materials, premises and equipment to the training and personal hygiene of staff. Detailed, written procedures are essential for each process that could affect the quality of the finished product. All materials, production environments and relevant documentation must be quality controlled, and there must be systems to provide documented proof that correct procedures are consistently followed at each step in the manufacturing process – every time a product is made.

The manufacture of ATMPs must be in accordance with GMP guidelines for human medicinal products, as laid down in Directive 2003/94/EC. This is designed to enhance the consistency and predicted efficacy of the clinical performance of ATMPs. Annex 2 of the EU Guidelines

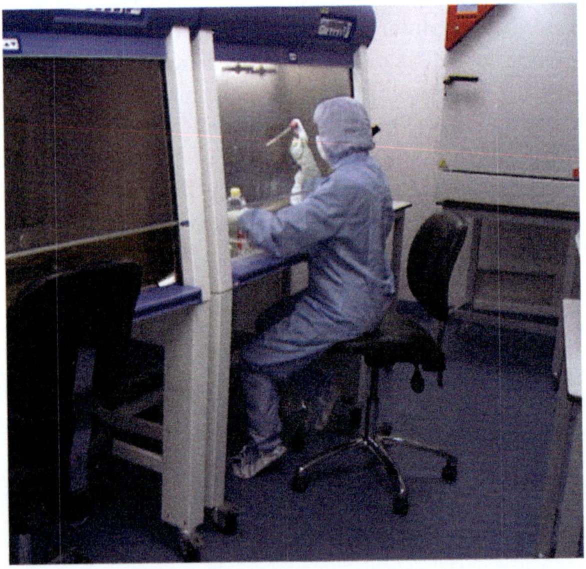

Figure 8.3 ATMP manufacturing in a licensed facility (photo taken in the GMP production rooms of the Newcastle Cellular Therapies Facility).

for Good Manufacturing Practices for Medicinal Products for Human and Veterinary Use (Eudralex, vol. 4) includes GMP procedures specific to ATMPs. Of great significance, the Annex recognises the inherent variability and increased risks of microbial contamination, as well as the transfer of pathogens associated with biological processes and materials. There must also be an understanding of the variability of human or animal cells and tissues as starting raw materials. Quality risk management principles are therefore critical in the development of control strategies seeking to diminish the impact of such biological variation. Quality risk management is therefore a core principle in the control of ATMP manufacture and is now incorporated into Annex 20 of the GMP guidelines.

8.3.5 European Union Tissue and Cells Directives (EUTCD)

The EUTCD provides a harmonised framework for the regulation of the quality and safety of human tissues and cells across Europe. There is clearly an overlap between the ATMP regulations and the EUTCD with respect to human tissues and cells. The EUTCD is comprised of the parent Directive 2004/23/EC and the two implementing technical Directives 2006/17/EC and 2006/86/EC which accompanied it. Specifically, these Directives set the standards for the quality and safety for the donation, procurement, testing, processing, preservation, storage and distribution of human tissues and cells. In addition, they provide a system for the traceability of tissues from donor to patient, including the requirements for tissue establishment accreditation for these activities. Directive 2004/23/EC came into force in April 2006 and transposed into national law by majority of the Member States in 2008. Where an ATMP contains human tissue or cells, the EUTCD applies in relation to the donation, procurement and testing of the starting human tissue and cells. Aspects such as processing to manufacture an ATMP, preservation, final product release, storage and distribution, are covered by the ATMP regulations. It is therefore the case that tiers of regulation may very well apply in the manufacture of ATMPs involving national authorities regulating activities under the EUTCD (such as the Human Tissue Authority in the UK), and competent authorities regulating ATMPs (such as the MHRA in the UK), overseen in the EU by the EMA. One of the earliest clinical studies of a cell-based therapy that sought to achieve Marketing Authorisation status in the UK was that of the Newcastle group incorporating the

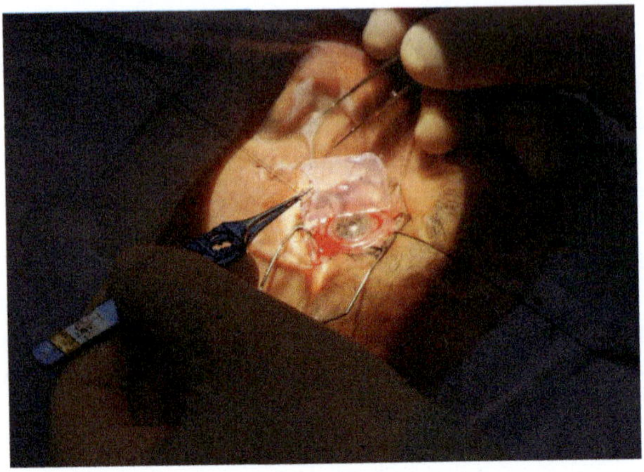

Figure 8.4 Application of a sheet of autologous limbal stem cells that have been expanded *ex vivo*, from a limbal explant, on a sheet of human amniotic membrane as a substrate, following superficial keratectomy (debridement) of corneal pannus scar tissue.

Department for Ophthalmology of the Newcastle, Royal Victoria Infirmary (RVI), the Institute of Genetic Medicine at Newcastle University, and the Human Stem Cell Laboratory at the RVI. The team led by Professor Francisco Figueiredo, Professor Majlinda Lako and Professor Anne Dickinson conducted Phase I and Phase II clinical trials of autologous limbal stem cells for the treatment of unilateral total limbal stem cell deficiency (LSCD), resulting from ocular surface burns (chemical or thermal), leaving patients with impaired vision, ocular surface inflammation and pain (Kolli et al., 2010).

8.4 ATMP Case Study: Autologous Limbal Stem Cell Therapy: the Newcastle Experience

Over the last 10 years there has been a steady increase in the clinical development of autologous cell products. The main aim of this case study is to share our experience and the difficulties of an investigator-led project, developing a new ATMP, and in advancing a cell therapy product in a regulatory environment in which both sponsor and regulators were still learning. This review focuses on how the existing regulations impacted on the development of an autologous cell therapy product to treat an orphan-designated disease, such as limbal stem cell deficiency (LSCD). The project was based on a centralised manufacturing and direct delivery clinical-site model (e.g. a clinical centre of excellence) using a pre-existing and experienced good manufacturing practice (GMP) academic manufacturing facility for bone-marrow cells to facilitate and expedite the development process of our new product (e.g. *ex vivo* expanded limbal epithelial stem cells).

As described earlier in this chapter, all manufacturing of medicinal products requires compliance with the standards of GMP to ensure their quality, safety and efficacy. In the UK, the Medicines and Healthcare products Regulatory Agency (MHRA) is the supervisory authority for UK manufacturers of centrally authorised ATMPs. To that effect, the MHRA is the competent authority for the assessment of applications for clinical trial authorisations (CTA) and the associated manufacturer's licence for investigational ATMPs.

In reality, our Newcastle Group broke new ground in the UK. There was no evidence of any previously registered clinical trials on limbal stem cells (LSC) in the UK prior to our UK Government (Medical Research Council) funded Phase II clinical trial. Our trial was to evaluate the safety and efficacy of cultured human autologous limbal epithelium for the treatment of unilateral total limbal stem cell deficiency. At the time, there was limited understanding regionally and nationally about what was required.

The Newcastle LSC research group is composed of a clinical team, support from academic specialists in the field, a GMP laboratory-based team, and the clinical trial was managed by the Newcastle Clinical Trials Unit, who provided a specific, named Clinical Trial Manager for each study. There is a monthly Trial Management Group meeting consisting of the clinical and management teams to discuss all aspects of the study to monitor progress. A Trial Oversight Committee meets 2 to 3 times a year and consists of a lay member serving as a patient representative, the clinical team and management team, an independent clinician, the trial statistician, an independent trial statistician and a representative from the funding body. The purpose of this committee is to primarily focus on pharmacovigilance and patient safety, but also to ensure the trial is progressing according to pre-set timelines and protocols and adhering to all regulations. The Clinical Trial Manager carries out scheduled study monitoring visits three times a year, to ensure accurate and timely documentation and consenting has taken place.

The clinical team is composed of a Consultant Ophthalmologist and a Clinical Research Fellow responsible for the recruitment, treatment and follow-up of all patients and for performing all associated tests and investigations during the course of the trial. The clinical team is also responsible for pharmacovigilance during the trial, with diligent reporting of every adverse event, serious or otherwise, related or unrelated to the investigational medicinal product (IMP), during the clinical trial. A data manager transfers the data captured in the study proformas into an electronic data capture system, developed specifically for the Newcastle LSC clinical study.

The GMP laboratory-based team is composed of a Director of the Newcastle Cellular Therapies Facility (NCTF), a Quality Control Manager, a Production Manager and a Qualified Person, as well as other technical staff within the facility. The Director oversees all production processes and projects under way at the NCTF. The manufacturing process occurred under the direct supervision of the Quality Control and Production Managers, who were responsible for ensuring and maintaining the quality of the manufacturing process and the cell therapy product. Under the MHRA licence, on completion of the manufacturing process, each batch of the IMP must be released for clinical use by a Qualified Person (QP), who must assess the release criteria and confirm full adherence to GMP regulations and guidelines for each individual manufacturing process. Each separate autologous manufacturing process constitutes a different batch of the product, and each batch is, by definition, manufactured for only one named patient.

The high frequency of patients presenting with ocular surface burns to the Corneal Service in the Ophthalmology Department of the Newcastle-upon-Tyne Hospitals NHS Foundation Trust led to the development of a Phase I clinical trial investigating the treatment of unilateral total limbal stem cell deficiency with *ex vivo*-expanded autologous limbal stem cells. This trial started in 2006, under the auspices of Directives 2004/23/EC and 2001/83/EC of the European Parliament and Council, regulated in the UK by the Human Tissue Authority (HTA). Due to the eventual clinical use of the product, the requirements for GMP compliance were considered important from the very beginning of our product development, with minimal transportation requirements given the proximity between the GMP lab and the clinical site.

Standards for the donation, procurement, testing, processing, preservation, storage and distribution of tissue and cells as described earlier in this chapter were set and overseen by the Human Tissue Authority (HTA). The HTA designates a Designated Individual (DI) to supervise the activity covered by the HTA licence. Due to the commitments of our clinical study in procuring, testing, processing, preserving, storing and distributing human tissue, our premises possessed an HTA licence and had access to a nominated DI.

In accordance with the regulatory requirements for a Phase I clinical trial involving Human subjects, we were required to obtain ethical approval for the conduct of the study prior to commencing recruitment for the trial. This was obtained from a regional Research Ethics Committee (REC) in December 2002. This committee consists of up to 18 members (a third of which are lay members), and meets at set intervals to review all applications before determining their suitability as a clinical trial. RECs are responsible for safeguarding the rights, dignity, safety and well-being of study participants. In addition, any clinical trial carried out within an NHS trust must be sponsored by the relevant NHS trust. Our clinical trials were sponsored by the Newcastle Hospitals NHS Foundation Trust.

In the UK, all new treatments available on the National Health Service (NHS) have to be previously approved by the National Institute for Health and Care Excellence (NICE), who determines if a new treatment is both evidence-based and cost-effective for the NHS. As our studies involved a research treatment that as a result had not yet gained approval from NICE, the provision of such a treatment had to be overseen by the Clinical Governance Department at the Newcastle Hospitals NHS Foundation Trust, who gave their approval for the clinical trials to take place.

On successful completion of our Phase I clinical trial, funding was sought to carry out a Phase II clinical trial using the same, established technique of autologous limbal stem cell culture. Our Phase I clinical trial took place from 2003 to 2009, whereby a system for growing human limbal epithelium without using animal cells or animal products was developed. The process was successfully validated in a GMP HTA-licensed laboratory (Human Stem Cell Laboratory, Haematological Sciences, Institute of Cellular Medicine, Newcastle University, UK). This subsequently led to the successful transplantation of 8 patients with unilateral total limbal stem cell deficiency (LSCD) with *ex vivo* expanded autologous limbal stem cells. In all 8 cases, symptom severity significantly declined ($p < 0.002$ for both pain and visual impairment from the LSCD). There was a significant improvement in Snellen visual

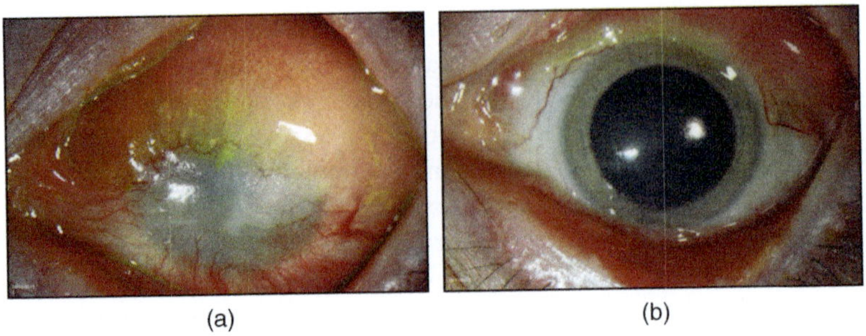

Figure 8.5 Colour photographs of patient with unilateral total limbal stem cell deficiency (a) before autologous limbal stem cell transplantation, and (b) 12 months after transplantation.

acuity in all 8 patients. In addition, corneal impression cytology at 6 months post-procedure has confirmed the reversal of conjunctivalisation (Kolli et al., 2010).

The Phase II trial design sought to include a further 24 patients (based on statistical power calculations), with a longer follow-up period of 36 months, as was required by the MHRA. The trial objective was to determine the safety and efficacy of the treatment proposed.

Approval was sought from the same necessary regulatory bodies as before to conduct a Phase II clinical trial for the use of *ex vivo*-expanded autologous limbal stem cells for transplantation for the treatment of unilateral total limbal stem cell deficiency. However, in the intervening period between the conduct of the Phase I trial and the application for the Phase II trial, the proposed treatment gained the classification of an advanced therapy investigational medicinal product (ATIMP). This new classification resulted from the incorporation of the ATMP regulations and was made on the basis that the cells were subjected to *substantial manipulation.*

The classification as an ATIMP necessarily required the involvement of the Medicines and Healthcare products Regulatory Agency (MHRA). Therefore, a Clinical Trial Authorisation (CTA) application and its associated documents had to be prepared and submitted in addition to seeking a GMP manufacturing licence. This regulatory burden led to a subsequent delay in starting the Phase II trial. Just as for our Phase I study, the Phase II study also required a favourable opinion from the REC (initially obtained in July 2007, then following re-classification of product as ATIMP was re-obtained in February

2012) before applying for a CTA (in December 2011; provisionally approved in December 2011 pending REC approval).

As stated, in the UK, the production of an ATIMP requires a Clinical Trial Authorisation (CTA) from the MHRA. A fee is due to the MHRA for each application (currently £3,400). In addition, a positive opinion from an ethics committee must also be obtained prior to starting a clinical trial in the UK. Each clinical trial must also be registered with the European Clinical Trials Database and obtain a EudraCT number. On its receipt, the MHRA will process the application according to the type of trial (A, B or C). For type A trials, the Clinical Trial Notification Scheme applies, where an acknowledgement of the application will be returned from the MHRA together with a note explaining that the trial may start 14 days from the receipt of the notification, if no objections have been raised. This notification scheme does not apply where a clinical trial is investigating a medicinal product, and each application must be fully considered by the MHRA. Our clinical trial fell into this category, and the MHRA provided an initial response within 30 days of receipt of a valid application (received on 19 December 2011, approved 21 December 2011).

Our application was for a Phase II clinical trial of an ATIMP, autologous limbal stem cells. With each application, a number of documents are required to be sent to the MHRA for inspection:

- covering letter
- clinical trial application form
- protocol document
- investigator's Brochure (IB)
- investigational Medicinal Product Dossier (IMPD)
- content of the labelling of the IMP
- proof of payment.

The Protocol document is carefully designed and describes the trial from start to finish, including how and where the study is to be conducted, a description of which subjects are eligible for inclusion, all the study processes and investigations, a schedule or timeline of follow-up and a description of the length of the study. The protocol may incorporate any standard operating procedures (SOPs) relating to the manufacturing process. It is important that the protocol document also highlights any relevant regulations applicable to the study and how the monitoring of safety relating to the treatment proposed and reporting of pharmacovigilance is to take place.

The Investigator's Brochure (IB) is a comprehensive document describing all information relating to the investigational product and is intended to assist the investigator in the conduct of the study and management of study subjects during the clinical trial. The purpose of the IB is to provide the principal investigator(s) with sufficient safety or performance data from pre-clinical investigation(s) and/or clinical investigation(s) to justify human exposure to the investigational device specified in the clinical investigational protocol. The IB shall be updated throughout the course of the clinical investigation as significant new information becomes available. The principal investigator(s) shall acknowledge the receipt of the IB and all subsequent amendments and keep all information confidential.

The Investigational Medicinal Product Dossier (IMPD) is a requirement in Europe and is based on the Directive 2001/20/EC of the European parliament and the council of 4 April 2001. When applying for a CTA, an IMPD is required when little or no information about an IMP has been previously submitted to a competent authority. The IMPD summarises information related to the quality, manufacture

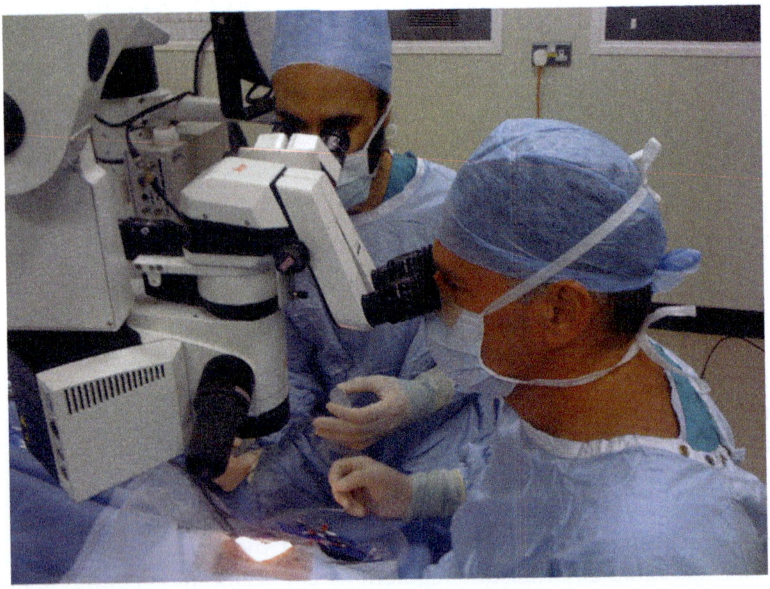

Figure 8.6 Implantation of the first limbal stem cell product in our Phase II clinical trial.

and control of any IMP and data from any previous clinical and non-clinical studies.

Recruitment for our Phase II clinical trial opened in February 2012 and in July 2012 the first ATIMP was released for implantation to the first patient in the trial.

The Phase II trial was overseen by the MHRA, who carry out regular inspections of the site at which manufacturing takes place. The clinical trial team is given prior notice of these inspections, and a GMP compliance report has to be completed by the clinical trial team prior to a scheduled inspection. During an inspection, the MHRA team will interview all relevant personnel, review relevant documents and conduct a site visit(s). At a closing meeting at the end of the inspection, the inspector will provide feedback and discuss any deficiencies identified and discuss timelines for corrective actions. Deficiencies are graded as critical, major or other. A "critical" deficiency is one that has produced or significantly risks producing a product that is harmful to humans or veterinary patients, or which could result in a harmful residue in a food-producing animal that could pose a risk to human consumption. A "major" deficiency is a non-critical deficiency that may produce a product that does not comply with its marketing authorisation, indicates a major deviation from GMP or the terms of the manufacturing licence, indicates a failure to carry out satisfactory batch-release procedures or a failure of the QP to fulfil their duties, or a combination of several "other" deficiencies, which on their own may not be major, but together may represent a major deficiency. An "other" deficiency is one which cannot be classified as "critical" or "major", but demonstrates a deviation from GMP.

Following the inspection, a post-inspection letter is issued, confirming any deficiencies identified, to which a response is required, outlining corrective actions and dates for when these will be completed. The inspector then reviews the response, and if they accept it, a GMP certificate is issued with the inspection report. If the inspector identifies any "critical" deficiencies, or that agreed actions from previous inspection deficiencies have not been resolved, they will contact the Inspection Action Group (IAG), who have the authority to refuse or suspend your GMP licence, increase inspection visits, or request a meeting with the licence holder. Our Phase II trial had such an MHRA inspection in April 2013, with no deficiencies identified and indeed, we received full praise from the inspector.

During the initial course of the Phase II clinical study, it became evident that some potentially eligible patients were being excluded from the study due to them having received a previous amniotic membrane transplant to the donor eye. However, in chemical injuries, often both eyes are affected, and both may receive an amniotic membrane transplant to aid healing of the ocular surface. However, as one eye may be less affected than the other, it may heal completely to an extent where there is no clinical or histological evidence of LSCD. Thus we determined that if a proposed donor eye had previously received amniotic membrane transplantation but was otherwise normal, with normal impression cytology, the patient would not need to be excluded from the study on that basis.

We had also initially proposed that previous participants of the study could not be re-recruited into the study. This would have excluded patients who had previously had a failed limbal biopsy, which had not resulted in an autologous limbal stem cell (ALSC) transplant. We requested to amend this to specify that only those who had received an ALSC transplant should be excluded, allowing a second limbal biopsy to be performed in willing patients. This change was backed up by the fact that other centres were already routinely performing two limbal biopsies, proving its safety due to its small size. A substantial amendment to the clinical trial has to be submitted and approved by both REC and the MHRA. There are slight variations in which documents each body requires to review, and these are clearly documented on their respective websites. Our substantial amendment was approved on 28 October 2013.

To date, the clinical experience of our trials has been reported at international conferences and data acquired published in high impact peer-reviewed journals. The therapy has permitted 32 patients throughout both clinical trials to receive autologous limbal stem cell transplantation, and 21 patients to subsequently receive a corneal transplant, restoring their sight. The principal clinical outcome restoration of the corneal epithelium has been achieved in all patients, with a concurrent improvement in patient-reported outcomes. No serious adverse events arising from the GMP manufacturing process, or indeed the ATIMP, have been reported in any patients. Our experience has been extremely informative both in terms of our clinical trial but also in terms of the practical application of ATMP regulations and GMP manufacturing processes. Our package of data was submitted to the CAT of the EMA with a view to seeking an initial opinion to confirm

ATMP classification, which was indeed received on 5 December 2014 (EMA/74581/2014). The CAT was also approached for Protocol Assistance with a view to providing a gap analysis on our data package so that upon submission of an MA application, all foreseeable issues had been addressed. In the end however, such is the competitiveness of the field of ATMP development, an alternative group in Italy, Chiesi Farmaceutici S.p.A. had applied for a MA for a similar product in March 2013, after being awarded orphan designation in November 2008. Chiesi Farmaceutici S.p.A. were conditionally awarded the MA in February 2015 for their product. Though this has some impact on future pathways to develop our product, which remains materially distinct from the Chiesi product, our experience has been extremely positive and our patients have benefitted immensely from the efforts of the whole team.

8.5 Conclusion

For the clinical development of ATMPs, the same basic requirements as for medicinal products generally apply. There is therefore a degree of conflict between the conventional medicinal product authorisation requirements, and the inherent diversity and variable biological properties of cells as raw materials, that are also reactive to variations in their environments. Cells are complex and react to their micro-environment in ways which can modulate cell viability, function, purity, stability, phenotype, potency, longevity, sterility and even tumourogenicity. In manufacturing an ATMP, all of these aspects that do not apply to non-biological pharmaceuticals, need to be addressed. The regulatory framework provided by the EMA for ATMPs can be seen to be mindful of the challenges that licensing ATMPs present. Investigations of pharmacokinetics, mechanisms of action, dose finding and clinical trials need be performed in accordance to Directive 2001/20/EC. Such classical studies may not be applicable with a proposed ATMP, and therefore require alternative approaches to establish at least a minimally effective and efficacious formulation.

When an ATMP for development through clinical trials contains human tissues or cells as a raw material, the requirements of the EUTCD must be met for donor consent, donor eligibility, data protection and confidentiality and ultimately, for traceability as required for the ATMP regulation, the EUTCD, and for Good Clinical Practice requirements. After treatment with an ATMP, robust systems of pharmacovigilance are needed as laid down in Directive 2001/20/EC, Regulation (EC) 726/2004 and 1394/2007 and the GCP guidelines for ATMPs. Normal concepts of a clinical study may prove particularly challenging in an ATMP trial, as clinical efficacy or safety might only be apparent after several years. The assessment of clinical outcomes and of associated long-term safety issues need to be considered, but these could be diverse, ranging from infections, immunogenicity and even the possibility of cell-based medicines leading to malignant transformations.

The analysis of efficacy and pharmacovigilance are crucial aspects in determining the success of clinical trials and such follow-up could extend beyond the end of the clinical trial to monitor specific safety issues such as tumourogenicity, teratogenicity, or the impact of adjunct therapies such as immunosuppression for allogeneic cell therapies. Of particular challenge is timely assessment to demonstrate compliance with serological criteria, for example screening for viral markers of HIV, Hepatitis B and C or human T-lymphotropic virus before commencing manufacture, and the screening for microbiological contamination during and after manufacture. Data on potential microbial contamination of a final product may not be available at the time of product release, and so the sterility of a product, or the risk it presents to the patient, may not be known at the time of treatment. Similarly, issues of product stability arise once cells are removed from the controlled environment of their processing into a final product presentation. These variables, which may present risks to the product, justify the development and validation of rapid test methods of potency and predicted efficacy specific to each therapy as it is developed. Clearly, therefore, as is a consistent recommendation of the authors of this chapter, early consultation with regulatory authorities such as the CAT is recommended to establish the clinical development programme.

References

Code of Federal Regulations, Title 21, vol. 8, revised 1 April 2014 (CITE: 21CFR1271.3). Part 1271 – Human Cells, Tissues, and Cellular and Tissue-based Products.

Commission Directive 2006/17/EC of 8 February 2006. implementing Directive 2004/23/EC of the European Parliament and of the Council as regards certain technical requirements for the donation, procurement and testing of human tissues and cells.

Council Directive 65/65/EEC of 26 January 1965 on the approximation of provisions laid down by Law, Regulation or Administrative Action relating to proprietary medicinal products.

Directive 2001/83/EC of the European Parliament and of the Council of 6 November 2001 on the Community Code Relating to Medicinal Products for Human Use.

Directive 2004/23/EC of the European Parliament and of the Council of 31 March 2004 on setting standards of quality and safety for the donation, procurement, testing, processing, preservation, storage and distribution of human tissues and cells.

check 2000 http://ec.europa.eu/health/documents/eudralex/index_en.htm

Human Tissue (Quality and Safety for Human Application) Regulations 2007). Available at: http://www.legislation.gov.uk/uksi/2007/1523/contents/made

Kolli, S., Ahmad, S., Lako, M. and Figueiredo, F.C. (2010) Successful clinical implementation of corneal epithelial stem cell therapy for treatment of unilateral limbal stem cell deficiency. *Stem Cells*, **28** (3), 597–610.

Qi, M., Barbaro, B., Wang, S. *et al.* (2009) Human pancreatic islet isolation. Part II: Purification and culture of human islets. *J. Vis. Exp.*, **27**, e1343.

Reflection paper on classification of advanced therapy medicinal products: EMA/CAT/600280/2010 Rev. Available at: http://www.ema.europa.eu/docs/en_GB/document_library/Scientific_guideline/2014/06/WC500169466.pdf

Regulation (EC) No. 141/2000 of the European Parliament and of the Council of 16 December 1999 on orphan medicinal products.

Regulation (EC) No. 726/2004 of the European Parliament and of the Council of 31 March 2004, laying down Community procedures for the authorisation and supervision of medicinal products for human and veterinary use and establishing a European Medicines Agency.

Regulation (EC) No. 1394/2007 of the European Parliament and of the Council of 13 November 2007 on advanced therapy medicinal products and amending Directive 2001/83/EC and Regulation (EC) No. 726/2004 (OJ L324, 10 December 2007, p. 121).

Rules and Guidance for Pharmaceutical Manufacturers and Distributors. Pharmaceutical Press, Jan 2015, ISBN 978 0 85711 171 5.

The Human Medicines Regulation 2012. SI 2012 No. 1916. Available at: http://www.legislation.gov.uk/uksi/2012/1916/pdfs/uksi_20121916_en.pdf

Index

Tables are denoted in **bold**: Figures and diagrams in *italic*

Abbreviation used in subheadings
ATMPs = advanced therapy medicinal products
LSCD = limbal stem cell deficiency
MSCs = mesenchymal stem cells
STB = stirred tank bioreactors

a

acoustophoretic separation techniques 135–137, *136*, **158**
active cell separation techniques 133
acute myocardial infarction, cell numbers required **71**
adherent cell therapy
- cell characterization assays **81**
- cell expansion 14, 15–18
- microcarriers 43–46, **44**

adipose tissue-derived stromal cells (ASCs) 47
advanced therapy investigational medicinal products (ATIMPs) 228–229, 231, 232
advanced therapy medicinal products (ATMPs) 4, 23, 211–212, 233–234
- classification of types 212–216, *217*
- LSCD example *224*, 224–233, *228*, *230*
- *see also* regulations/legislature

Advanced Tissue Sciences (ATS) 3
affinity-based separation techniques 142–144, *143*, **158**
agitation rates/agitators 59
- cell expansion/recovery 94–96, **95**, 103, 106–111, **107–108**, 119–120
- STB process development 38, 41, 43, 59
- *see also* impellers

allogeneic stem cell therapy 1, 2, 5
- bioreactor vs. automation 93
- cell characterization 65, 70
- clinical trials 8, *10*, 188

Bioprocessing for Cell Based Therapies, First Edition. Edited by Che J. Connon.

allogeneic stem cell therapy (*contd.*)
  manufacturing operations 17, 26
  separation techniques 128–129
  STB process development 35
  storage, short-term 189, *192*, 193
  see also mesenchymal stem cells
Alternative Tangential Flow (ATF) system 20
American Association of Tissue Banks 180
ammonia, medium exchange 54–55, *57*, 97, *99*, 100
animal materials, microcarriers 45, 49
antifoam-C 47
antioxidants, hypothermic biopreservation solutions 194
Apligraf 2, 3, 4, 26, 188, 191, 204
apoptosis, cryopreservation 171, 189–190, 197
aqueous two-phase system (ATPS) separation techniques 137–138, **158**
Argos therapeutic cancer immunotherapy 30
ASCs (adipose tissue-derived stromal cells) 47
Aseptic Technologies, LX1 filling line 26
assays
  cell characterization 66–68, **81**
  cryopreserved cell viability 178
  STBs 46–47
  see also cell surface markers
Asymptote, VIA Freeze technology 27, 176
ATF (Alternative Tangential Flow) system 20
ATMPs. *see* advanced therapy medicinal products
ATPS (aqueous two-phase system) separation techniques 137–138, **158**
ATS (Advanced Tissue Sciences) 3
autologous (donor) serum (AS) 74
autologous stem cell therapy 2, 5
  bioreactor vs. automation 93
  cell characterization 65, 70
  clinical trials 8, *10*, 188
  separation techniques 120, 133
  STB process development 35
  storage, short-term 189, 191–193, *192*
  see also ChondroCelect; limbal stem cell deficiency; Provenge®
automated processes
  centrifuge systems, blood component processing, 139–140, **158**
  future developments 28, 29–30
  cell expansion/recovery 93–96, **93**, 120

b

bacterial contamination 14, 51, 223, 234
batch centrifugation, manufacturing operations 21
batch processing, separation techniques 134–135

biobanking systems 5, *6*, 13, *14*, 71, 181
BioCEP cell enrichment process 145
biologics, definitions 4
biology. *see* cell biology
biomarkers. *see* cell surface markers
biopharmaceuticals 5, 6, 23, 93–94, 103
bioreactor technology 18–19, 26. *see also* stirred tank bioreactors
  vs. automation, for cell expansion/recovery **93**, 93–96
blood component processing, automated centrifuge systems 139–140
bone marrow stem cells 66
  case study, expansion in STBs 59–60
  cell characterization 73
  hypothermic biopreservation solutions 195–197
  leukaemia patients 1–2
  licensed products 8
  manufacturing operations 23
  see also mesenchymal stem cells
bovine serum. *see* fetal bovine serum
bubbles, fluid mechanics 95–96, 109
burns 188. *see also* limbal stem cell deficiency

C

calcium alginate storage media 198–199, **200**, 203
cancer
  potential for causation 92, 128, 234
  therapies 10–11, 30
cardiac applications, cell numbers required **71**
Carticel® 188, 190–191
caspase 9 independent programmed cell death, cryopreservation 171
CAT (Committee for Advanced Therapies) 219, 221
CCE (Centrifugal Counterflow Elutriation) 138–139, **158**
cell adhesion. *see* adherent cell therapy
cell banks 5, *6*, 13, *14*, 71, 181
cell biology
  assays 46–47, **81**
  cell types 45–46
  and scale-up 94–95, **95**
cell characterization, for cryopreservation 170–172, *173*. *see also* cell phenotype comparability testing
cell death (apoptosis), cryopreservation 171, 189–190, 197
Cell Enrichment Process (CEP), BioCEP 145
cell expansion 13–19, 46, 79
cell expansion/recovery, MSCs 91–92, 120–121
  bioreactor vs. automation **93**, 93–96, **95**, 120
  harvest 117–120, *120*, 121
  large-scale expansion in STBs 111–117, *115*, *116*
  microcarrier screening 103–108, **104–105**, **107**, **108**

cell expansion/recovery, MSCs (*contd.*)
small scale operations 96–103, **98**, **99**, **101–102**
spinner flask culture 108–111, *110*, **112–113**, 114, 118
cell homing assays 67, 70, **81**, 82
Cell Medicas Cytovir (CMV), clinical trials 13
cell membranes
cryopreservation 170–171, 172, 178
hypothermic biopreservation solutions 197–198
cell numbers. *see* growth kinetics
cell pausing (suspended animation) 194
cell phenotype comparability testing 65–66
cell numbers required **71**
challenges 72–75
characterization assays **81**
future developments 80–83
International Society for Cellular Therapy criteria 75–77
MultiStem® Allogeneic Cell Therapy Product 69–71
omics screens 78–80
potency assays 77–78
product consistency in development 66–69, **68**
cell recovery. *see* cell expansion/recovery
cell seeding density, small scale STB operations 51–52, *52*, *53*
cell separation. *see* separation techniques
CellStack® plates 47–49, **48**
cell storage. *see* storage
cell surface markers
cell characterization 66–68, 73, 76, **81**, 82–83
STB process development 56, *58*
cell therapies overview 1–3, 31
administration to patients 27–28
classification of types 9
clinical trials 8–11, *10*
companies, products and procedures 8
definitions 2, 4
formulation, fill and finish 24–27
future developments 28–31
licensed products 7–8
scale of manufacture 7
see also advanced therapy medicinal products; manufacturing operations; regulations/legislature
Cell Therapy Catapult 11
cell therapy products 187. *see also* advanced therapy medicinal products
cell variability. *see* cell phenotype comparability testing; homogeneity
cell viability
cryopreservation 171–172, 174, 177–178, 182
storage, short-term 189, 190, **199–201**
Centrifugal Counterflow Elutriation (CCE) 138–139, **158**
centrifugation 13, 21–23, 138–140, **158**

CEP (Cell Enrichment Process) 145
chemokine ligand 5 (CXCL5) 78
Chiesi Farmaceutici S.p.A. 233
Chinese Hamster Ovary cells (CHO) 45–46, 92, 202
CHMP (Committee for Medicinal Products for Human Use) 218
ChondroCelect 12, 65, 78, 188, 190, 204
chondrocyte-derived products 7
chromatography, optical sorting 151, *154*
chromosomal duplications/deletions 92
classifications, cell therapy *9*, 212–216, *217*
cleaning processes 14, 20–22. *see also* separation techniques
clean rooms 28–29
clinical trials 8–11, *10*, 13, 188
  ATMPs 233–234
  LSCD example 225, 227–230, *230*, 232–234
  regulations/legislature 216
  STB process development 35
clinician-led approaches 1–2, *9*, *10*
CliniMACs Magnetic-Activated Cell Sorting technology 24, 130, *132*, 157
closed processing
  future developments 29, 30
  separation techniques 134, 135
CMV (Cell Medicas Cytovir), clinical trials 13
COBE 2991 Cell Processor system 22, 23, 71, 139–140, 157
cold shock, mesenchymal stromal cells 197. *see also* temperature
collagen-coated microcarriers 72, 73, **104**
colligative action theory, cryopreservation 168
combined advanced therapy medicinal products, definition 213
Commission Directive 65/66/EEC/1 (1965) 218
Committee for Advanced Therapies (CAT) 219, 221
Committee for Medicinal Products for Human Use (CHMP) 218
Common Technical Document (CTD) 218
communication issues, engineers and biologists 94
Compact SelecT T-flasks 16, 117
company-led approaches 2, 8, *9*
comparability testing. *see* cell phenotype comparability testing
concentration, manufacturing operations 20–22
consistency. *see* homogeneity
contamination, microbial 14, 51, 223, 234
continuous centrifugation 21
continuous processing, separation techniques 134–135, 147
controlled rate freezers (CRFs), cryopreserved cells 179–180

cooling protocols, cryopreservation 170–172, 174–176
cord blood derived haematopoietic progenitor cells 7
corneal biopsies 15, 224, 226. *see also* limbal stem cell deficiency
Corning CellStack® plates 47–49, **48**
cost factors 2–3, 35
  financial incentives 219
  separation techniques 129
  STB process development 47–49, **48**
CPAs (cryoprotective agents) 168–172, *173*, 174, *175*, 189–190
CRFs (controlled rate freezers), cryopreserved cells 179–180
critical quality attributes (CQAs) 69, 70, 91–92, 94, 96
Crohn's disease, clinical trials 188
cross-flow filters 155–156, **158**
cryopreservation 27, 167–168, 181–182
  cell characterization 75, 170–172, *173*
  cell survival vs. cooling rates *175*
  development challenges 169–178
  large-scale freezing 178
  manufacturing challenges 179–181
  storage, short-term 189
CTAs (clinical trial authorisations) 225, 228–229, 230, 232–234
cryoprotective agents (CPAs) 168–172, *173*, 174, *175*, 189–190
CTPs. *see* cell therapy products
Cultispher microcarriers **104**, 114
Cytodex 3 microcarriers **104**, 108, **112–113**, 114, 119
cytokines, cell characterization 78, **81**
Cytori's Celution 189

d

DEAE-Sephadex beads 111
definitions
  advanced therapy medicinal products 212–214
  cell therapies 2, 4
  Good Manufacturing Practice 221–222
delivery systems. *see* shipment
dendritic cells (DC), gene expression profiling 79
Dermagraft 2, 3, 4, 188, 191
detachment, MSCs 118, 119. *see also* harvest; separation techniques
deterministic lateral displacement (DLD), separation techniques 141, **158**
dextran, separation techniques 137
Dickinson, Professor Anne 224
dielectrophoresis (DEP), separation techniques 140–141, **158**

differentiation
  assays 47, **81**
  cell characterization 73
  manufacturing operations 19–20
  STB process development 56, *58*
dimethyl sulfoxide (DMSO) 168–169, *173*, 175, 189
dipole magnetic flow fractionation (DMF), cell separation 145
dissociation, MSCs 118, 119. *see also* harvest; separation techniques
disposable materials, manufacturing operations 14
distributed manufacture, future developments 30
DLD (deterministic lateral displacement), separation techniques 141, **158**
DMF (dipole magnetic flow fractionation), cell separation 145
DMSO (dimethyl sulfoxide) 168–169, *173*, 175, 189
doubling times, cell numbers. *see* growth kinetics
downstream processing, manufacturing operations 11, *11*, 20–24, 31

e

economic viability, separation techniques 129. *see also* cost factors
elephant ear segmented impellers 37, *37*
embedded integrated processing, separation techniques 133
embryonic stem cells (ESC) 46
  clinical trials 8
  genetic engineering 141–142
  microcarriers 73
  storage 203
EMD Millipore 37, 38, 42, 43, 47, 48, 51
energy landscapes, optical sorting 151, 152, 154, *155*
engineering, and scale-up 94–95, **95**, 97
Epicel 188, 191, 204. *see also* Genzyme
epigenetic profiling. *see* gene expression profiling
ESC. *see* embryonic stem cells
ethics 12, 227
EudraLex regulations 218
European Medicines Agency (EMA) 214, 218, 220–221, 223, 232, 233
  European Regulations 212, 214, 216–219, 226, 233, 234
    Committee for Advanced Therapies 221
    Good Manufacturing Practice 220, 221–223, *222*
    Hospital Exemption rules/ Specials procedures 219–220
    Orphan Medicinal Product Designation 220–221
    Tissue and Cells Directives 215, 216, 219, 223–224, 233
expansion. *see* cell expansion

f

FDA. *see* Food and Drug Administration
feeding strategies, MSCs 54–55
fetal bovine serum (FBS) 65
 cell characterization 80
 moving away from (serum-free applications) 74–75
 STB process development 60–61
 storage, short-term 189
fibrin microbeads 198
fibroblast-based products 7, 9, 109, 148
Figueiredo, Professor Francisco 224
fill, formulation and finish 25–27
filtration, manufacturing operations 21–22. *see also* separation techniques
finish, formulation and fill 25–27
5 L bioreactors *115*, 117
FLC3000, Bosch 26
Fleming, Sir Alexander 92
flow cytometric fluorescent-activated cell sorting (FACS) 130, *131*, 157
flow cytometry assays 46–47, **81**
fluid mechanics, and scale-up 95–96, 109
foetal bovine serum. *see* fetal bovine serum
Food and Drug Administration (FDA) 2, 3, 4, 7, 9, 11, 24
 clinical trials 188
 Code of Federal Regulations 213
 potency testing 77
formulation, fill and finish 25–27

g

gelatin-based microcarriers 45
GE (manufacturer), microcarriers **44**, 45
gene expression profiling
 cell phenotype comparability testing 68, 78, 79–80, **81**
 cryopreservation 169
 STB process development 60
gene ontology analysis 79
gene therapy medicinal products 212
genetic engineering, separation techniques 141–142, **158**
Genzyme 17, 188, 190–191
glass state vitrification 169, *173*
glass-transition temperature (Tg) 172, 176, 177
glucose, medium exchange 54–55, *57*, 97, 99, **101–102**, **107–108**
glutamine, feeding strategies, MSCs 54–55, *57*
Good Clinical Practice (GCP), ATMPs 233–234
Good Manufacturing Practice (GMP) 4
 cryopreserved cells 179
 LSCD example 225, 226, 227, 231, 232
 magnetic-activated cell sorting 24
 regulations/legislature 212, 219, 220, 221–223, *222*
 separation techniques 128, 134, 157
 storage, short-term 189, 190, 191, 193, 195, 205
graft-versus-host disease (GVHD) 35, **71**, 75, 188

growth factors, cell characterization 82
growth kinetics
cell characterization **68**, **71**,73
cell expansion/recovery *98*, *99*, 100, **101**, **102**, 108, **112–113**, 117
separation techniques 129
STB process development 46, 47, 54–55, *56*
GvHD (graft-versus-host disease) 35, **71**, 75, 188

h

haematopoietic progenitor cells (HPCs) 7–8, 203–204
harvest 20–22, 114, 117–120, *120*, 121. *see also* separation techniques
HDF (hydrodynamic filtration), separation techniques 142, **158**
heat transfer, stirred tank bioreactors 42–43
Herceptin 92
hESCs. *see* embryonic stem cells
high-grade clean rooms 28
high-gradient magnetic separation (HGMS) 148
high resolution separation techniques 134
Hillex microcarriers **44**, 50, **105**
hMSCs. *see* mesenchymal stem cells
hollow fibre reactors, cell expansion 16–17
homing of cells, assays 67, 70, **81**, 82
homogeneity, product 66–69, **68**, 92, 127. *see also* cell phenotype comparability testing
Hospital Exemption (HE) rules, ATMPs 219–224
HPCs (haematopoietic progenitor cells) 7–8, 203–204
HPL (human platelet lysate) 61, 74
HSA storage medium 195–196, **199–201**
HTA (Human Tissue Authority) 215–216, 226, 227
human adipose-derived MSCs (hASCs) 196
human embryonic stem cells. *see* embryonic stem cells
human mesenchymal stem cells. *see* mesenchymal stem cells
human platelet lysate (HPL) 61, 74
human progenitor cells *9*
Human Tissue Act (2004) 215, 216
Human Tissue Authority (HTA) 215–216, 226, 227
Human Tissue (Quality and Safety for Human application) Regulations (2007) 215
human umbilical vein endothelial cells (HUVECs) 77
hydrodynamic filtration (HDF), separation techniques 142, **158**
hypothermic biopreservation 187, 191–194, *192*
cell therapy applications 204–205

hypothermic biopreservation (*contd.*)
  mesenchymal stem cells 194–202, **199–201**
  optimal temperatures for storage 203–204
HypoThermosol-FRS® storage medium 196–197, 204
hypoxia, MSCs 54, 97, 100

i
IB (Investigator's Brochure), LSCD example 230
ice-crystals, cryopreservation 171, 172, *173*, 174
ICH (International Conference on Harmonisation) 218
immortalised cells, teratomas 92, 128
immunoadsorption separation techniques 142–144, *143*, **158**
immunomagnetic cell separation techniques 145, *146*, **158**
immunosuppression, cell characterization **81**, 82
IMPD (Investigational Medicinal Product Dossier) 230–231
impellers
  cell expansion/recovery 103, 109, 110, *110*, 111
  STB process development *37*, 37–38, 40
induced pluripotent stem cells (iPSC) 46, 80
inertial migration, separation techniques 145–147, **158**
Information Standard for Blood and Transplant (ISBT) 128, 180
Inspection Action Group (IAG), LSCD example 231
integrated processing, separation techniques 133
intellectual property rights, cell homogeneity 67
interleukin 8 (IL-8) 78
International Conference on Harmonisation (ICH) 218
International Society for Cellular Therapy (ISCT) 49, 75–77
intracellular ice, cryopreservation 167, 168, 172, 174
Investigational Medicinal Product Dossier (IMPD) 230–231
investigative assessments 68. *see also* assays
Investigator's Brochure (IB), LSCD example 230
Invetech, closed processing 30
iPSC (induced pluripotent stem cells) 46, 80
ISBT (Information Standard for Blood and Transplant) 128, 180
ischemic stroke 70, **71**
ISCT (International Society for Cellular Therapy) 49, 75–77

k
keratinocyte-based products 7
knee cartilage repair 188
Kolmogorov eddy length, STBs 40
kSep system, KBI 23

l

label-free separation techniques 130–133, **158**
  magnetic cell sorting 147–150, *149*
  optical sorting 151–154, *152*, *153*, *154*, *155*
lactic acid, medium exchange 54–55, *57*, 97, *99*, 100, **101**, **102**, **107–108**
Lako, Professor Majlinda 224
large-scale expansion. *see* scale-up
large-scale freezing, cryopreservation 178
legislature. *see* European regulations; regulations/legislature
leukaemia patients 1
licensed products 7–8
limbal stem cell deficiency (LSCD) *224*, 224–233, *228*, *230*
low resolution separation techniques 133–134
LX1 filling line, Aseptic Technologies 26
lyophilization 26

m

MACI (matrix applied characterized autologous cultured chondrocytes) 17
Magisterial Formula 219
magnetic-activated cell sorting (MACS) 24, 130, *132*, 157
magnetic cell sorting, label-free separation 147–150, *149*
magnetic separation, manufacturing operations 24
magnetic solution-based separation techniques 148–150, *149*, **158**
magnetophoretic cell separation techniques 147–148, *149*, **158**
malignant transformations, pharmacovigilance 234. *see also* teratomas
mammalian cell separation. *see* separation techniques
mammalian cell storage. *see* storage
manufacturing authorisation. *see* marketing authorisation
manufacturing operations overview 11, *11*
  cell banks *14*
  cell expansion 13–19
  cell separation 13
  challenges 5, 7, 29
  downstream processing 11, *11*, 20–24, 31
  sourcing 12–13
  upstream processing *11*, 11, 13–20, *14*, 31
  see also specific processes
MAPC (multipotent adult progenitor cells), characterization 69–71, 76
marine impellers *37*, 38
markers. *see* cell surface markers
marketing authorisation (MA), LSCD example 216, 218, 221, 223, 233
mass balance 97
master cell banks 5, *6*, 13, *14*
Matrigel coatings, microcarriers 73, 74, 78, **81**

matrix applied characterized autologous cultured chondrocytes (MACI) 17
Mazur's two phase hypothesis, cryopreservation *175*, 175
Medicines and Healthcare products Regulatory Agency (MHRA) 216, 219, 225, 226, 228, 229, 231
medium exchange, waste products/nutrients 20, 93, 97, *98*, *99*, 100, **101**, **102**, 103
medium scale operations, STB process development 55–59, *56*, *57*, *58*
membranes, cell. *see* cell membranes
mesenchymal progenitor cells (MPCs) 17
mesenchymal stem cells (MSCs)
  advantages of using 194–195
  cell characterization 66, 73–76, 78, 80, 82
  clinical trials 8, 188
  hypothermic biopreservation solutions 194–202, **199–201**
  see also bone marrow stem cells; cell expansion/recovery, hMSCs
methylation screening 79. *see also* gene expression profiling
MHRA. *see* Medicines and Healthcare products Regulatory Agency
microbial contamination 14, 51, 223, 234
microcarriers
  cell characterization 72–74, 78, **81**
  cell expansion/recovery 103–108, **104–105**, **107**, **108**, 111, 114
  small scale STB operations 52–53, *54*
  STB process development 35–36, *39*, 43–45, **44**, **48**, 49–50, *50*, 59, 94
microfilters 150–151
microscale laminar vortices 150, **158**
microscale membranes 155–156, **158**
migration of cells, assays 67, 70, **81**, 82
Miltenyi Biotech 24
mixing phenomena, STBs *37*, 37–40, *39*
Mobius® CellReady bioreactors **48**, 48, 55, *56*
MPCs (mesenchymal progenitor cells) 17
mRNA profiling, cell characterization 79, **81**
MSCs. *see* mesenchymal stem cells
multi-layer reactors, cell expansion 15–16
multi-orifice flow fractionation (MOFF) devices 147
multipotent adult progenitor cells (MAPC), characterization 69–71, 76
Multistem cell therapy product 17, 69–71, 75, 78
myocardial infarction, cell numbers required **71**

n

National Institute for Health and Care Excellence (NICE) 227
necrosis, cryopreservation 171, 189–190
network solidification, cryopreservation 178
Newcastle Cellular Therapies Facility *222*, *224*, 224–233, *228*, *230*
next-generation sequencing **81**
normal flow filter (NNF) devices 150–151
Nunc Cell Factories, comparisons with other systems 47–49, **48**

o

ocular surface burns 224, 226. *see also* limbal stem cell deficiency
off-the-shelf cell therapies 133
omics screens, cell characterization 78–80
one donor-one recipient therapies 1, 2
  bone marrow stem cells 23
  cancer therapies 11
  manufacture 5, 14, 15, 24, 26
  sourcing the cells 12, 13
one to many therapies 2
  manufacture 5, 14, 15
  sourcing the cells 12–13
open systems, separation techniques 134
optical sorting, label-free 151–154, *152*, *153*, *154*, *155*, **158**
Orange Guide (Rules and Guidance for Pharmaceutical Manufacturers and Distributors) 212
  Orphan Medicinal Product Designation 220–221
osmotic stress, cryopreservation 170–172
oxygen, dissolved
  cell expansion/recovery 97, *99*, 100, **101**, 114, 117
  STB process development 54, 55
oxygen transfer rate (kLa), stirred tank bioreactors 40–42, **42**

p

packaging
  cryopreserved cells 179
  manufacturing operations 25–26
pancreatic islets cells 214–215, *215*
Parkinson's disease 82
passage 15
  cell characterization 67, 75, 82
  cell expansion/recovery 97, 101, **101**, **102**, 103, 106, 118
  STB process development 49, 61
patient specific cell therapies 133
patients
  cell therapy administration 27–28
  consent, cell sourcing 12
PBS storage medium 195, **199–201**
PDGF (platelet-derived growth factor) 82

PEG (polyethylene glycol) 137
perfusion cell culture 19, 20
peripheral blood mononuclear cells (PBMCs) 77–78
pharmacovigilance 234. *see also* safety
pH gradients, STB process development 53, 55, 114, 117
phenotype comparability testing. *see* cell phenotype comparability testing
pillar-type filters 156–157, **158**
pitched blade impellers 37, *37*
plastic microcarriers 15, 18, 76
  cell expansion/recovery **105**, 108, 114, 119
  STB process development **44**, 49
platelet-derived growth factor (PDGF) 82
pluriBead immunoadsorption *143*, 144
pluripotent stem cells 46. *see also* embryonic stem cells; induced pluripotent stem cells
Pluronic F-68 surfactant 47
Pneumatics, Unifuge system 23
polyethylene glycol (PEG) 137
polystyrene microcarrier 45
potency 67, 77–78 , 80, **81**
power input per volume, impeller design 38–39
preservation 13, 25–27, 75. *see also* cryopreservation; storage
process development, stirred tank bioreactors 50–55
process monitoring targets, stirred tank bioreactors 46
Prochymal® 65, 188
products/procedures 8, 9, *10*
  consistency. *see* homogeneity
  monitoring targets, STBs 46
  scale-up programmes 94
progressive solidification, cryopreservation 178
proliferative capacity, manufacturing operations 13, 14
prostate cancer 11, 188
Provenge® 11, 26, 65, 188
purification 22. *see also* separation techniques
PVA microcarriers **105**

q

quadrupole magnetic cell sorters (QMS) 145, *146*
Quantum Cell Expansion System 22, 23, 71
Quantum hollow fibre reactors 17

r

radiation exposure during shipment, cryopreserved cells 180–181
recombinant protein-coated microcarriers 46, 103, **104**
recovery, cryopreserved cells 177–178
Refine Technology 20
regulations/legislature 4, 211–216, 233–234
  classification of types 212–216, *217*
  cryopreserved cells 181

Hospital Exemption rules/ Specials procedures 219–224
LSCD example *224*, 224–233, *228*, *230*
see also European Regulations
REMEDiE project 5
Research and Development programmes, scale-up 94
Research Ethics Committee (REC) 227
risk. *see* safety
robotics. *see* automated processes
rocking platforms, bioreactor technology 19
Rules and Guidance for Pharmaceutical Manufacturers and Distributors (Pharmaceutical Press) 212
Rushton impeller *37*, 37

S

safety 223
microbial contamination 14, 51, 223, 234
risk and benefit to patients 1
screening 234
separation techniques 128
see also Good Manufacturing Practice
SanBio Inc. 82
Sartorius Stedim Biotech Group 16, 37–38, 42
scaffold technology, cell expansion 17–18
scale of manufacture 7, 21, 29, 30
scale-out. *see* automated processes
scale-up
hMSCs 93, 94, 117, 119, 121
separation techniques 130–135
STB process development 36, 49
storage, short-term 193
see also cell expansion/recovery; cell phenotype comparability testing
scientific investigative assessments 68. *see also* assays
screening, product safety 234
seeding density, small scale STB operations 51–52, *52*, *53*
separation techniques 13, 127–129, 157–159, **158**
acoustophoretic sorting 135–137, *136*
aqueous two-phase systems 137–138
centrifugation 138–140
deterministic lateral displacement 141
dielectrophoresis 140–141
flow cytometric fluorescent-activated cell sorting 130, *131*, 157
genetic engineering 141–142
hMSCs 118
hydrodynamic filtration 142
immunoadsorption 142–144, *143*
immunomagnetic cell separation 145, *146*
inertial migration 145–147
magnetic cell sorting 130, *132*, 145, 147–150, *149*, 157
manufacturing operations 22–24

separation techniques (*contd.*)
  microscale laminar vortices 150
  normal flow filter devices 150–151
  optical sorting 151–154, *152*, *153*, *154*, *155*
  pillar and weir-type filters 156–157
  scale-up 130–135
  tangential flow filters 155–156
  see also centrifugation; harvest
Sepax centrifugal cell processing system 13
serum-free applications 74–75
shear protectants 47
shear stress 19, 21–23, 25
  and scale-up 72, 95, **95**
shelf life 2, 25, 26, 27
shipment 26–27
  cryopreserved cells 177, 180–181
Sigma (manufacturer) **44**, 45
skin biopsies 15
skin-equivalent products 2, 3
small and medium-sized enterprises (SMEs), financial incentives 219
small scale operations, stirred tank bioreactors 50–55, *52*
solid suspension types, stirred tank bioreactors *39*
somatic cell therapy medicinal products 213
somatic stem cells, clinical trials 8
sourcing the cells, manufacturing operations 12–13
sparging methodology 40–42
special needs test 219
specials procedures, ATMPs 219–224
spinal cord injury 82
spinner flasks 40, 108–111, *110*, **112–113**, 114, 118
split flow thin fractionation (SPLITT) 145
standardisation, separation techniques 129
standard operating procedures (SOPs), LSCD example 229
standards, industry 29, 128, 180
stem cell properties, cell characterization 72–74
Steriflip® vacuum filter *120*
Stirling engine cooling 27
stirred tank bioreactors (STB) 18–19, 35–36
  vs. automated systems **93**, 93–96
  biopharmaceuticals 93–94
  case study for expansion of bone marrow derived MSCs 59–60
  cell biology 45–47
  cell seeding density *52*, *53*
  comparisons with other systems 47–49, **48**
  future developments 60–61
  heat transfer 42–43
  hMSCs 111–117, *115*, *116*
  impeller types *37*
  medium scale operations 55–59, *56*, *57*, *58*
  microcarriers 43–45, **44**, **48**, 49–50, *50*, 59
  mixing phenomena 37–40
  oxygen transfer rate 40–42, **42**

small scale operations 50–55, *52*
solid suspension types *39*
storage 187, 204, 205
cell characterization 75
cryopreserved cells 176–177
hypothermic biopreservation solutions 195–199, **199–201**
logistics 188–193, *192*
manufacturing operations 13, 25–27
optimal temperatures 202–204
see also cryopreservation
stroke, cell numbers required **71**
substantial manipulations 4
definitions 213, 214
LSCD example 228
supply chains
cryopreserved cells 180–181
manufacturing operations 11, 12, 25, 26–27, 30
storage, short-term 188–189
surface markers. *see* cell surface markers
surface tension, cell characterization 73
surrogate potency assays 78
suspension cell types 45–46
suspension expansion 14, 18–19
Synthemax **44**, 61, 72, **105**

t

TACSI PL system 139, 157
tangential flow filters (TFF) 155–156, **158**
temperature
glass-transition 172, 176, 177
harvest 119
preservation and storage 27, 202–204
see also cryopreservation; hypothermic biopreservation
teratomas, immortalised cells 92, 128
terminology 2
Terumo BCT system 22, 23, 71, 139–140, 157
T-flasks 15–16, 93–96, 106, 111, 117
TGFbeta (transforming growth factor) 82
thalidomide tragedy 218
thawing/recovery, cryopreserved cells 177–178
The Orange Guide (Rules and Guidance for Pharmaceutical Manufacturers and Distributors) 212
three-dimensional (3D) processes 15, 17, 171, 195
cell characterization 72, 78
separation 137, 141, *152*, 154, *155*
tissue engineered products 213
tissue expansion 14, 15. *see also* cell expansion/recovery
toxicity, cryoprotectant agents (CPAs) 168, 169, 171, 172, 174, *175*, 189–190
transportation. *see* shipment
trypsin-EDTA 119
turbulence, stirred tank bioreactors 109, 110, *110*
tweezers, optical 151, *152*, *153*
two cell therapy products 2

two-dimensional (2D) processes
cell characterization 65, 71, 73
stirred tank bioreactors 35, 38, **42**, 47, 49, 50, 60
two phase hypothesis, cryopreservation *175*, 175

u

Unifuge system 23
University of Winconsin (UW) organ preservation solution 194
upscaling. *see* scale-up
upstream processing 11, *11*, 13–20, *14*, 31
US FDA. *see* Food and Drug Administration

v

vaccine development 111, 114
variability. *see* homogeneity
vascular endothelial growth factor (VEGF) 78, 82
viability. *see* cell viability
VIAFreeze technology 27, 176
Viaspan® preservation solution 194
vitrification 169, 170, 171, *173*

w

washing, manufacturing operations 20–22. *see also* separation techniques
wave bioreactor technology 19
weir-type separation techniques 156–157, **158**
World Health Organization (WHO) 212

x

xeno-free microcarriers 61, 74–75, **104**
xenogeneic serum 189

y

yield. *see* growth kinetics

Printed and bound by CPI Group (UK) Ltd, Croydon, CR0 4YY
16/09/2026
14963472-0001